ICELAND

BLE ROUTE

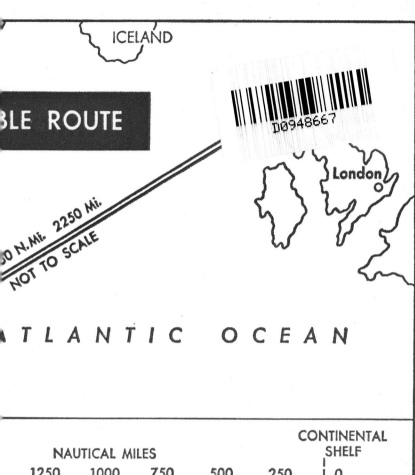

50 N.Mi. 2250 Mi.

NOT TO SCALE

London

ATLANTIC OCEAN

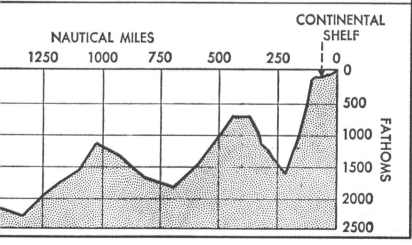

NAUTICAL MILES

CONTINENTAL SHELF

| 1250 | 1000 | 750 | 500 | 250 | 0 |

| 0 |
| 500 |
| 1000 | FATHOMS |
| 1500 |
| 2000 |
| 2500 |

VOICE ACROSS THE SEA

By the Same Author

VOICE ACROSS THE SEA

Arthur C. Clarke

William Luscombe Publisher Limited
in association with Mitchell Beazley

First published in 1958
This completely revised
edition published by
William Luscombe Publisher
Limited
Artists House
14 Manette Street
London WIV 5LB
1974

ISBN o 86002 0681

Set and printed in Great Britain by
Butler & Tanner Ltd, Frome and
London

Contents

To John Pierce
who suggested this book
and bullied me into writing it

Illustrations

Preface to the Second Edition

This book was written at the instigation and with the encouragement of Dr. John Pierce, soon after the laying of the first transatlantic telephone cable (TAT 1) in 1956. Two months before it was completed, the Space Age dawned, and I ended the final chapter with these predictions:

> The first relay satellites may be nothing more than radio mirrors circling round the world at fairly low altitudes. They might be balloons a hundred or so feet in diameter, covered with metallic paint and inflated by gas pressure when they attained the desired height. . . . Even to the naked eye, they would be quite conspicuous objects, being the brightest stars in the dawn and sunset sky.
>
> If the reflecting spheres were at a reasonably low altitude – say five hundred to a thousand miles – they would move across the sky so rapidly that it would be difficult to track them. . . . It would be necessary to have a whole string of reflectors, so that as one sank below the horizon another rose into sight.
>
> The answer to this problem is to move the reflectors further away from earth so that they travel more slowly. Indeed, there is an orbit, twenty-two thousand miles above the Equator, where they would not move at all but would appear fixed in the sky. . . . Three satellites, equidistantly spaced round the Earth, could provide complete radio and TV coverage over the whole globe. . . . The satellites needed to provide a global service would not, however, be merely passive reflectors. They would have to be true relay stations, amplifying and rebroadcasting the programs they received from earth . . . it seems more than likely that before the transatlantic cable has reached its twenty year expectation of life, we will be able to make concrete plans for the system of world-girdling satellites.

The only thing wrong with this prophecy was the time-table. The 100-foot ECHO was launched into its thousand-mile-high orbit in 1960, and the first triad of synchronous satellites completed the global system in 1969 – not, as I suggested, around 1976 . . .

So now we have two competing – and co-operating – systems of transoceanic communications. Far from downgrading the importance of the cables, satellites have acted as a stimulus. New cables are being laid at this very moment; while I was working on this revision, the eyes of the world were focused on the deepest sea rescue in history, when the minisub Pisces III, with two men aboard, was safely retrieved from 1,565 feet. She had been on charter to the British Post Office, and had been digging a trench at the bottom of the Atlantic to bury a new telephone cable out of the reach of trawlers.*

To do justice to the astonishing developments of the last sixteen years would demand another book at least as long as this one, and I have not attempted such a task. What I have done is to leave the historical material (Chapters 1–22) untouched, except for necessary up-datings. I have completely re-written Chapter 23, 'The New Cables', to summarise developments since the laying of TAT 1. And I have written two entirely new chapters, 24 and 25, to deal with communications satellites – which, I believe, are now beginning to re-structure the patterns of human society. This subject is also discussed in more detail in my book *Voices From The Sky*.

Despite the glimpses of the future with which the book ends its main emphasis is still upon the heroic efforts of the great Victorians to bridge the Atlantic. The laying of the first cables was the Apollo Project of its day, the *Great Eastern* an engineering achievement worthy to be set beside the Saturn v. There are lessons here of the utmost relevance to our own time, when we are faced with technological decisions involving billions of dollars, and affecting the future of whole nations – indeed, of the entire human race. We will need, in the years ahead, precisely the same

---

* The communications satellites have earned much revenue, handling the traffic while broken cables were being repaired. But I never really believed Dr. Joseph Charyk, President of COMSAT, when he once confessed to me: 'Of course, you realise those trawlers belong to *us* . . .'

courage and foresight which our great-grandfathers brought to bear upon the problems of their age.

But perhaps the most important moral to be drawn from this first great conquest of space is one which many have now forgotten, and which therefore needs reasserting as never before.

Technology may be possible without Civilisation; but Civilisation is not possible without Technology.

Colombo,
Sri Lanka,
1974.

CHAPTER 1

Introduction

This is the story of Man's newest victory in an age-old conflict – his war against the sea. It is a story of great moral courage, of scientific skill, of million-dollar gambles; and though it affects every one of us directly or indirectly, it is almost entirely unknown to the general public.

Our civilisation could not exist without efficient communications; we find it impossible to imagine a time when it took a month to get a message across the Atlantic and another month (if the winds were favourable) to receive the reply. It is hard to see how international trade or cultural exchanges could flourish or even exist in such circumstances. News from far parts of the world must have been rather like the information that astronomers glean about distant stars – something that happened a long time ago, and about which there is nothing that can be done.

This state of affairs has existed for the greater part of human history. When Queen Victoria came to the throne she had no swifter means of sending messages to the far parts of her empire than had Julius Caesar – or, for that matter, than Moses. It is true that semaphore systems had been invented, in which letters were indicated by the position of moving arms like those of old-fashioned railway signals, but they were clumsy and of limited application. The galloping horse and the wind-driven sailing ship remained the swiftest means of transport, as they had been for five thousand years.

Not until the scientists of the early nineteenth century started to investigate the curious properties of electricity was a servant discovered which within little more than two lifetimes would change the face of the world and sweep away the ancient barriers of time and distance. It was soon found that the 'electric fluid' travelled through conducting wires at a velocity so great that there was no way of measuring it, and at once ingenious experimenters in many countries attempted to use this fact for the transmission of messages. By 1840 the electric telegraph had left the laboratory and become a commercial instrument of vast possibilities. Within ten years it had covered most of Europe and the settled portions of North America – but it still stopped at the edge of the sea.

How the ocean was at last defeated is the main theme of this book. In 1858 a handful of far-sighted men succeeded in laying a telegraph cable across the North Atlantic, and at the closing of a switch the gap between Europe and America dwindled abruptly from a month to a second.

But this first triumph was short-lived; the ocean was too strong to be bound by so slender a thread, and in a few days the continents were as far apart as ever. The way in which, after an eight-year saga of almost unbelievable courage and persistence, a successful Atlantic telegraph was finally laid is one of the great engineering epics of all time, and has many lessons for us even today.

The Victorians built well; some of the cables laid down in the last century are still in use today, after having carried incomputable millions of words for mankind. There is a section of cable in mid-Atlantic which began work in 1873 and has been quietly doing its job while the theologians agonised over Darwin, the Curies discovered radium, a couple of bicycle mechanics in North Carolina fitted an engine to an oversized kite, Einstein gave up his job at the patent office, Fermi piled up uranium blocks in a Chicago squash court, and the first rocket climbed into space. It would be hard to find any other technical device which has given continuous service while the world around it has changed to such an extent.

The submarine cable, however, has one fundamental limitation. It can transmit telegraph signals but it cannot – except over relatively short distances – carry the far more complex pattern of vibrations which constitutes speech. Graham Bell's invention of the telephone in 1876 opened up a new era in communications, but had no effect on the whole-wide submarine cable system. The requirements of speech transmission were so severe that there seemed no hope at all of ever sending the human voice across the Atlantic.

The discovery of radio changed the situation radically, and also presented the submarine cables with a major challenge. To the great surprise of science, and the great good fortune of the communications industry, it turned out that the earth is surrounded by an invisible mirror which reflects back radio waves which would otherwise escape into space. When this mirror – the ionosphere – is co-operating, it is possible to send speech around the curve of the globe after one or more reflections. Unfortunately the ionosphere is not a smooth, stable layer; it is continually changing under the influence of the sun, and during times of solar disturbance it may be so convulsed that long-distance radio is impossible. Even when conditions are good, radio links which depend on the ionosphere are liable to pick up all sorts of curious cracklings and bangings, for the universe is a very noisy place in the radio spectrum. Pascal, who complained that the silence of infinite space terrified him, was a little wide of the mark. He would have been astonished to learn that it is full of the sound of solar eruptions, exploding stars, and even colliding galaxies. These electromagnetic noises add a background, and all too often a foreground, to the radio messages transmitted from one continent to another.

Nevertheless, a radio-telephone service was established across the Atlantic in February 1927; until 1956, this was the only means whereby the human voice could pass from Europe to America Yet it is probably true to say that most people who have ever thought of the matter assumed that the transatlantic telephone depended on cables, not on radio. One German spy even claimed to have listened-in to conversations between Roosevelt and Churchill by tapping submarine cables; unfortunately for the truth of this story, Roosevelt had been dead for a dozen years before men spoke to each other across the bed of the Atlantic.

In 1956, the impossible was achieved and the first submarine telephone cable was laid between Europe and America. The inflexible laws which state that one cannot send speech for more than a few score miles through an underwater cable had not been repealed; they had been by-passed by a new and daring approach to the problem – one that involved sinking a chain of more than a hundred amplifiers, each more complex than the average radio set, along the bed of the ocean.

Any great engineering achievement, especially if it has long been considered impossible, can provide both an emotional and an intellectual stimulus. It is true that a submarine cable is not something that everyone can see, like a giant bridge, a skyscraper or an ocean liner. It does its work in the darkness of the abyss, in an unimaginable world of eternal night and cold and pressure, peopled by creatures which no man could have conceived in the wildest delirium. Yet it serves a function as vital as that of the nerves in the human body; it is an essential part of the world's communication system – which, if it ever failed, would throw us back instantly to the isolation of our ancestors.

From the nature of the subject, this book falls into two distinct sections. The first part is the more romantic, for it covers the brave pioneering days when fortunes were made and lost in bold gambles against the forces of Nature, and the fabulous *Great Eastern* dominated the seas as no ship will ever do again. By contrast, today's story is one of scientific, not physical, adventure; yet it will, I hope, appeal to those who have no technical background or interests whatsoever.

I would also like to emphasise that this is not a history of submarine communications. As far as it goes it is, I believe, accurate, but it makes no attempt to be complete. My object has been, frankly, to entertain as much as to instruct, and as a result I have wandered down some odd by-ways whenever the scenery has intrigued me. It will contribute little to anyone's understanding of telegraphy to know how Oliver Heaviside made tea, why Lord Kelvin's monocle revolutionised electrical measurements, what a Kentucky colonel was doing in Whitehall, how Western Union lost $3,000,000 in Alaska, and what unlikely articles the Victorians made from gutta-percha. Yet it is precisely such trivia that make history three-dimensional, and I do not apologise for including them.

Finally, there is one lesson above all which I hope this story will teach. From the beginning, the development of transatlantic communications has been an example of the way in which men of different nations could work together for a common objective, and the first telephone cable has been a classic case of such international co-operation. It was conceived and designed jointly by the American Telephone and Telegraph Company and the British Post Office, all ideas being pooled with complete freedom, and the best answers chosen quite irrespective of their source. The '*We* invented this – it must be best' philosophy was not allowed to intrude.

Now that men can at last speak across the Atlantic with perfect clarity and freedom from interference, a new era has opened which will profoundly affect the political, commercial and social relationships between Europe and America. To mention only one possibility which exists: for the first time musical concerts, plays, discussions and indeed the entire gamut of radio entertainment can now be exchanged 'live', without any loss of quality. The radio networks of the two continents can be linked together with a clarity never possible in the days when the link itself was a radio one. Today, when you call a friend on the far side of the Atlantic, you seem to be talking quietly together in the same room – not, as was too often the case in the old days, shouting over a high wind during a thunderstorm. There is a sense of presence, of almost physical contact, which has to be experienced to be appreciated.

The importance of this, and its effect upon cultural exchanges on all levels, can hardly be over-estimated. For today the future of the world, and perhaps the survival of mankind, depends upon the bonds which link together the Atlantic Powers.

CHAPTER 2

The Coming of the Telegraph

Like most great inventions, the electric telegraph has a complex and disputed ancestry. America, Russia, Germany and England have approximately equal claims to its origin, and although today Samuel Morse is remembered above most of his rivals, he was very far from being the first man to transmit information by electricity.

Morse sent his famous message 'What hath God wrought?' (a question which, incidentally, still lacks a suitable reply) on May 24, 1844. But one standard history of the subject lists no less than forty-seven telegraph systems between the years 1753 and 1839, and although most of these were only paper proposals, some of them actually worked.

Perhaps the first really determined attempt at the electrical communication of intelligence was Sömmering's 'chemical telegraph', constructed in Munich in 1809. In this system, every letter was represented by a separate wire which terminated at the bottom of a water-filled container. When current was passed through a given wire, bubbles formed at its end, and by watching where the bubbles appeared an observer could tell which letter was being transmitted. Although the method was hardly practical, it was a notable achievement and attracted much attention at the time.

A still more elaborate system, depending on static electricity, was set up in 1816 by Sir Francis Ronald in his garden at Hammersmith, London. Ronald erected no less than eight miles of overhead wire, and read messages passed through the line by the movement of light pithballs at the end of it. As they were electrified, their mutual repulsion swung them aside to expose the letter it was desired to transmit.

Sir Francis deserves credit as the first man to realise the possible business, social and international possibilities of this new method of communication. A brochure he published in 1823 was the first work on telegraphy ever printed; it even contained proposals for locating the position of faults on a telegraph line. Unfortunately, Sir Francis was about a generation too early. When he offered his system to the British Admiralty, he was told that Their Lordships were perfectly satisfied with the telegraph they already had, and there was no question of its being replaced by anything else. The Navy's 'telegraph' at that time consisted of a string of semaphore towers by which, in clear weather, messages could be transmitted from Portsmouth to London slightly faster than a pony express could have done the job.

By one of the minor ironies of technology, the Secretary of the Admiralty who signed the letter of rejection lived to write the article on telegraphy in the *Encyclopædia Britannica;* by another, Sir Francis's house was later occupied by William Morris, leader of the romantic, back-to-the-Middle-Ages revival, who could hardly have felt a great deal of sympathy for an invention which was to do so much to sweep mankind into a strange and tumultuous future.

The systems devised by Ronald, Sömmering and other inventors were impracticable because they lacked a simple and sensitive means of detecting the flow of electricity. In 1820, however, came the great discovery which was to make the world we know. The Danish scientist Oersted found that an electric current could produce a deflection of a magnet placed near it. For the first time, electricity had exerted force. From that simple observation stemmed in due course the myriads of generators, motors, relays, telephones, meters, loudspeakers and other electromagnetic devices which are now civilisation's most ubiquitous slaves.

Around 1825, this new knowledge was applied to telegraphy by Baron Schilling, attaché to the Russian Embassy in Munich, who had been much impressed by Sömmering's earlier work. Amongst other arrangements, Schilling devised a magnetic telegraph in which letters were indicated by movements of a needle over the white or black segments of a card. He employed a code based on the same principle as that later made famous by Morse; in Schilling's alphabet, A was 'black, white', B was 'black, black, black', C was 'black, white, white' and so on. (Such bi-signal alphabets, incidentally, go back to at least Greek and Roman times.)

Here at last was the basis of a really practical telegraph, and the time was ripe for its exploitation, which occurred almost simultaneously in America and England. In 1836. W. F. Cooke, an English medical student at Heidelberg, heard of Schilling's work, realised its importance, and immediately abandoned his intended profession. He knew a good thing when he saw one, and hurried back to England to find an electrical expert who could help him to put his ideas into practice, since his own knowledge of science was rudimentary.

The man he contacted was Charles Wheatstone, Professor of Physics at King's College, London. Wheatstone's name is remembered through a whole series of basic electrical inventions, the most famous being the Wheatstone bridge, a method of measuring resistances by balancing an unknown against a known one. I have some affection for his shade as a result of spending two years in the Wheatstone Laboratory at King's College, a period during which, at least according to the experiments recorded in my Practical Physics notebooks, the constants of Nature were remarkably variable.

Cooke and Wheatstone produced their first telegraph patent in June 1837, and carried out their first practical trials in the same year over a mile-and-a-quarter long line between two London railway stations. The receivers they used were the so-called needle instruments, in which letters were indicated by the deflection to right or left of vertical pointers. The system was slow and somewhat elaborate, but messages could be sent and read by unskilled staff. Instruments of this general type were still in use in out-of-the-way British railway stations well into the twentieth century.

For a long time railways and telegraphs went hand in hand; the new means of transport could not have operated without some such rapid form of communication. Within a very few years the steel rails and the copper wires had spread their networks over much of Europe, and Cooke and Wheatstone netted fortunes in royalties. Success promptly ruined their relationship, which broke up into a squalid argument as to which of them *really* invented the telegraph. The answer, of course, was neither.

While this was going on in England, a mildly talented portrait painter named Samuel Finley Breese Morse was trying, without much success, to get support for his ideas on the other side of the Atlantic. He had heard about the possibilities of electrical communication during a casual conversation with a fellow passenger while sailing back to the United States from Europe in 1832, and the concept had immediately fired his mind. Owing to the problem of making a living, however, he did not produce his first working telegraph instrument until 1836.

There is a striking parallel between the histories of Morse in America and Cooke in England. Each was an amateur scientist and each had to consult a professional in order to make any progress. Morse was helped by Joseph Henry, the great pioneer of electromagnetism, who has now given his name to the unit of induction; and in due time Morse and Henry became involved in quarrels over priority, exactly as Cooke and Wheatstone had done.

The beauty of Morse's system was its simplicity. It is so simple, indeed, that we tend to take it for granted and forget that someone had to invent it. Earlier telegraph systems had involved many wires and cumbersome sending and receiving apparatus. Morse produced a telegraph that needed only one wire (the earth providing a return circuit) and whose transmitter was nothing more than a key to make and break the connection. By means of the dot-dash code, this single key could send any letter or combination of letters.

The first receiver Morse built consisted of a magnet-operated pen which automatically wrote the incoming dots and dashes as jiggles on a moving tape, thus providing a permanent record. Very soon, however, it was discovered that the ear could interpret the long and short buzzes, and the Morse sounder or buzzer came into general use. It survives, virtually unchanged, to this day.

Morse was also responsible for introducing the relay, which, in theory at least, made it possible to transmit messages for indefinite distances.* In this simple but basic device the feeble current at the end of a telegraph line was used to close a contact which was, in effect, a second Morse key, starting a new current from another set of batteries along the next section of line. The

* There is strong evidence that Henry had invented the relay before Morse applied it.

relay was the earliest form of 'repeater', a device which we shall meet later in a much more sophisticated form.

After years of effort and a fruitless journey to Europe to sell his invention, Morse finally obtained $30,000 from Congress in 1842 for the construction of a line between Washington and Baltimore. The debate over the allocation does not reflect much credit on the elected representatives of the American people; several of them were quite unable to appreciate the difference between magnetism and mesmerism. But Morse got his money, and two years later America got the telegraph. Without it, the immense continent could never have become a united country.

The manner in which the telegraph spread from Atlantic to Pacific, the wars between the desperately competing companies, the eventual triumph of Western Union over its rivals – this is now part of American history and, indeed, folk-lore. For half a century, until he was displaced by automatic instruments such as the teleprinter, the telegraph operator was one of the picturesque, and essential, figures of the American scene. For a brief span of time he perfected and practised a skill which had lain dormant and unsuspected in mankind since the beginning of history: the ability to read and transmit up to forty words a minute by an almost continuous series of broken buzzes, hour after hour. Some of the feats of these men – of whom the young Edison was the most famous though certainly not the most characteristic – were quite incredible and could probably not be repeated today.

There is one well-authenticated account of a telegraph operator who, to show off, deliberately ignored his racing Morse sounder while it went full speed for a couple of minutes, and then sat down to catch up on the messages that had already come through. After fifteen minutes of writing, he eventually absorbed the backlog and was taking down the words at the moment they arrived. Such a feat of memory can only be compared to playing a dozen games of blindfold chess simultaneously – and against the clock.

Such skills have now vanished from the earth, because they are no longer needed. The Morse code – or its equivalent – still flashes around our planet at a rate which must be measured in millions of words a day. But it has now become a language understood and spoken almost entirely by machines, and not by men.

CHAPTER 3

Channel Crossing

By 1850 the tentacles of the electric telegraph had spread all over England, as well as over much of Europe and the more settled areas of North America. But the wires still stopped at the edge of the sea, and it was obvious where the first submarine cable should be laid – across the Straits of Dover.

The first serious scheme for a cross-Channel telegraph had been put forward by Professor Wheatstone in 1840 to a House of Commons Committee. A few years later he carried out experiments in Swansea Bay, Wales, sending signals between a boat and a lighthouse. These were not, however, the first underwater signals ever transmitted; priority for this seems to go to a Dr. O'Shaughnessy, Director of the East India Company's Telegraphs, who laid a primitive submarine cable across the River Hooghly in 1839. A little later – in 1842 – Morse carried out experiments in New York Harbour, sending signals through a length of rubber-insulated wire enclosed in a lead pipe. Although this is going ahead of our story, these tests had led Morse to conclude, as early as 1843, that 'telegraph communication may with certainty be established across the Atlantic Ocean. Startling as this may now seem, I am confident the time will come when the project will be realised.'

By an unlikely turn of events, the man, who first linked England and France together was a retired antique dealer. John Watkins Brett had made a fortune in this peculiar trade, and at the age of forty-five was still full of energy and prepared to try something new. His younger brother Jacob, who was an engineer, first interested him in the possibilities of submarine telegraphy, and between them the Bretts formed a grandiloquently titled 'General Oceanic and Subterranean Electric Printing Telegraph Company'. Perhaps it should be explained that it was the telegraphy, not the 'Electric Printing', which was intended to be subterranean.

After negotiating with the French Government, the Bretts

secured a ten-year concession for the laying of a cross-Channel cable, and contracted with the Gutta Percha Company for its manufacture. As was all too often the case with such pioneering projects, the whole scheme was rushed through far too quickly, without a proper understanding of the problems involved. The Bretts were working against a deadline; if they could not establish communication between France and England by September 1, 1850, their concession would lapse.

The cable was so primitive that it seems incredible that anyone could have expected it to work. It was nothing more than a single copper wire, surrounded by a quarter of an inch of gutta-percha for insulation. It was assumed that once the cable had been safely laid on the sea-bed nothing could happen to it and so it would need no armouring. Only the shore ends were given protection by being encased in lead tubes.

Very few people took the scheme seriously, and as usual those who knew least about it were the most critical. One gentleman, seeing the cable being paid out, declared roundly that the promoters must be fools; anyone should know that it was impossible to drag a wire that long over the rough bed of the Channel. He was under the impression that signals would be transmitted by jerking the cable, as in the system of wires, pulleys and bells which the wealthier Victorians employed to summon their numerous domestics from kitchen to drawing-room.

The Bretts had a scant three days to spare when they loaded their twenty-five miles of cable aboard the small steam tug *Goliah* and set out from Dover on the morning of August 28, 1850. The cable had been coiled on a large drum seven feet in diameter and fifteen feet long, which was placed on the after-deck with its axis horizontal, so that it spanned the entire width of the tiny boat. The drum, looking like an enormous cotton-reel, revolved as the cable was paid out, so as the *Goliah* proceeded she was rather like an angler letting his line run out by walking backwards away from a point where the hook had been fixed. This system was only practical with very small and light cables; all later ones were coiled in circular wells and paid out layer by layer.

Since the Bretts' cable was much too light to sink properly, it was necessary to attach lead weights to it every hundred yards or so. Despite the general chaos caused by this operation, and the strain on the cable when the *Goliah* stopped to fix the weights, the end of the cable was safely landed at Cap Gris-Nez on the evening of the 28th after an uneventful crossing.

There was great excitement as the automatic printer was connected up and the party on the French coast waited for the first message to come through – a flowery greeting from John Brett to Prince Louis Napoleon Bonaparte. Alas, all that emerged from the printer was a mass of jumbled characters which made no sense whatsoever; it almost appeared as if the English operators had been celebrating a little too soon. The automatic printer was disconnected and a needle instrument put in circuit; this time some words got through without mutilation, so at least the Bretts were able to claim that they had fulfilled the terms of their contract. But whether any complete and intelligible messages were exchanged seems highly doubtful, for the signals in both directions were equally jumbled.

They did not know it yet, but the telegraph engineers had now come face to face with an enemy who was to cause them endless trouble in the years ahead. At first sight, it would appear that if a properly insulated cable worked on land, it would work just as well in the sea. But this is not the case; when it is submerged in water, and thus surrounded by a conducting medium, the transmitting properties of a cable are completely altered. As we shall see later, it becomes much more sluggish owing to its greatly increased electrical capacity. Signals no longer pass through it at speeds comparable to that of light, but may move so slowly that before a 'dash' has emerged from the far end, a 'dot' may already be treading on its heels. It was this retardation that had foiled the Bretts. If their operators had slowed down their normal rate of sending to match the characteristics of the cable, the messages would have got through.

Unfortunately, there was no chance of continuing the experiments. When the tired and rather dispirited telegraphists sat down at their instruments the next morning, the line was completely dead. Electrical tests showed that it had broken somewhere near the French coast, and it was subsequently discovered that a fisherman had fouled the line with his anchor. As the line was so light he was able to haul it aboard, and he was immensely puzzled by this new kind of seaweed with a metal core. Thinking that it might be gold, he cut out a section to show his friends, and thus started the long war between the cable companies and the other users of the sea that has lasted to this day. More damage has been done to submarine cables by dragged anchors or trawls than by any other cause, and the annoyance is often mutual. A small boat that hooks its anchor around a modern armoured cable is as likely to lose its anchor as to damage the cable.

Despite its failure, the 1850 cable had shown that signals could be sent across the Channel. Nevertheless, the Bretts had the utmost difficulty in raising money for a second attempt, and it was not until a year later that the enterprise went ahead once more. This time the prime mover was Thomas Crampton, a railway engineer, who not only subscribed half the £15,000 needed for the project, but designed the new cable himself.

And this time it was a real cable, not a single insulated wire. The four gutta-percha insulated conductors produced by the Gutta Percha Company were protected with hemp, and a layer of galvanised iron was spun on top of that to act as armouring. No fisherman would be able to haul up this cable; it looked like a large hawser and weighed more than thirty times as much as its predecessor.

This very weight almost defeated the project when the cable was laid on September 25, 1851. The year before, it had been necessary to attach lead weights to the line to make it sink – but this cable was only too eager to reach the sea-bed. It paid out so fast that the inadequate brake could not prevent excessive wastage, and as the ship was also carried off-course by wind and tide the French coast was still a mile away when the end of the cable was reached. Luckily there was a spare length aboard for just such an emergency and a temporary splice was made to complete the connection. After a few weeks of testing, the cable was opened to the public, and no point in Europe was more than a few seconds away from England.

After the initial failure and the complete scepticism of all but a few enthusiasts, the establishment of this cross-Channel link – the world's first efficient submarine cable – created a great impression. With typically Victorian optimism, this new miracle of communications was hailed as a triumph for peace, which would undoubtedly improve the understanding and co-operation between nations. Today we are sadly aware that though civilisation cannot function without such links, it by no means follows that they automatically bring peace. As the mathematicians would say, they are necessary – but not sufficient.

Punch, that ubiquitous commentator upon the times, celebrated the event with a cartoon showing what appears to be a two-headed angel with an olive branch tripping lightly along the bed of the Channel, delicately balancing on the cable with one toe like a ballet dancer walking the tight-rope. According to the artist, the bottom of the English Channel is a much more interesting place than I have ever found it to be; it is liberally strewn with cutlasses, guns, broken spars and the skulls of unfortunate seamen.

The submarine cable boom was now under way; within the next two years the Gutta Percha Company, which had a virtual monopoly of core-insulation, supplied no less than 1,500 miles of covered wire to the manufacturers who provided the protective armouring. If one could have watched a map of Europe which showed, like an animated cartoon, the progress of the cables, the period 1851 to 1856 would have shown remarkable, and all too often fruitless, activity. Thin black lines would have extended out in all directions from England, only to fade away again after a short period of existence. There were two attempts to span the Irish Sea before a permanent cable was laid; then Dover and Ostend were successfully linked, and after that no less than four sound cables were laid between England and Holland.

In 1855, a famous cable was laid across the Black Sea for the British Government to speed communications in the Crimean War. (So much for peace and understanding between nations!) This cable was needed in such a hurry that there was no time to get it armoured; like the first cross-Channel cable, it was simply an insulated wire. Yet it gave good service for nearly a year, and helped to shorten a war which the incompetence of the general staff had done so much to prolong.

The Mediterranean was first tackled in 1854, when a short cable was successfully laid between Corsica and Sardinia. Then a longer link was established between Corsica and the Italian coast, but the telegraph engineers were now running – literally – into deep waters, and disastrous failures resulted from attempts to connect Sardinia and Algeria, so that Europe and Africa could speak to one another. These failures must have been tragically disheartening to all concerned, but their causes – as we shall see in Chapter 8 – now appear extremely comic.

The general principles of submarine telegraphy were being learned by trial and error; though they did not appreciate the privilege, the shareholders were paying for the education of their engineers. And as soon as a few submarine cables had been successfully laid, it was inevitable that men's thoughts would turn to spanning the most important ocean of all – the Atlantic.

With her overseas possessions, maritime interests and technical 'know-how' – primitive though that appears to us today – it was also inevitable that England should be the pioneer in the field of submarine cables, and not surprising that she has held this lead for a hundred years. Indeed, more than 90 per cent of the cables in the world have been made by a single British firm, the Telegraph Construction and Maintenance Company. Yet the initiative and drive which finally resulted in the laying of a successful Atlantic cable, after years of setbacks and disasters, came from an American.

It is time to meet Cyrus W. Field.

CHAPTER 4

A Great American

Cyrus West Field was one of the greatest Americans of the nineteenth century, but today there can be few of his countrymen who remember him. He opened up no frontiers, killed no Indians, founded no industrial empires, won no battles; the work he did has been buried deep in the Atlantic ooze for almost a hundred years. Yet he helped to change history, and now that his dream of a telegraph to Europe has been surpassed by a still more wonderful achievement, it is only right that we should pay tribute to the almost superhuman courage which enabled him to triumph over repeated disasters.

His face is looking at me now, across the century that lies between us. It is not at all the face of the international financier or the company promoter, though Field was both these things. The thin, sensitive nose, the regular features, the deep-set, brooding eyes – these add up to a poet or musician, not to the stereotyped sad success, indistinguishable from all his ulcer-ridden colleagues, we see today in the business section of *Time* magazine. 'Visionary and chivalrous' were the words applied to Field many years later, and no one without vision would have set off on the long and arduous quest that dominated his life for almost twelve years. But the vision would not have been enough without the practical shrewdness which had made him what would be a millionaire by our standards while he was still in his early thirties.

Cyrus Field was born on November 30, 1819, from New England stock, being a descendant of one Zechariah Field, who had emigrated from England around 1629. His father was a Congregational minister at Stockbridge, Mass., and perhaps because he was the youngest of seven sons Cyrus matured unusually early. When he was only fifteen he asked permission to leave home and seek his fortune; with $8 in his pocket he drove fifty miles to the Hudson and sailed downstream to New York.

Like many boys before and since, he discovered that the

streets of Manhattan were not as well paved with gold as he had hoped. During his first year as an errand boy in a Broadway dry-goods store he earned a dollar a week; though this was doubled in the second year, Cyrus came to the conclusion that his talents were not appreciated in New York and returned to Massachusetts. At the age of eighteen he became an assistant to his brother Matthew, a paper-maker, and only two years later went into business himself in the same trade.

Soon he was all set to make his fortune; he became a partner in a large New York firm of wholesale paper dealers, and – still only twenty-one – married Mary Bryan Stone, of Guilford, Conn.

Six months later, the roof fell in on him. The firm with which he had associated himself failed, and though he was the junior partner he was left holding the debts. Out of the wreckage he built Cyrus W. Field & Co., and worked with such intensity that his family saw him only on Sundays. By the time he was thirty-three he had paid off all his obligations, and was able to retire with $250,000 in the bank – all of which he had made in nine exhausting years.

Now he could relax; indeed, his doctor ordered him to do so. Like every other wealthy American, he 'did' Europe with his wife; then, rather more adventurously, he explored South America with his friend Frederick E. Church, a famous landscape painter of the time, whose study of that totally impossible subject, Niagara Falls, is still considered one of the best ever put on canvas. Field and Church crossed the Andes – no mean feat in those days – and brought back as souvenirs a live jaguar and an Indian boy. It would be interesting to know which gave them more trouble.

Field might have remained in obscure retirement for the rest of his days if chance had not brought him into contact with F. N. Gisborne, an English engineer engaged on building a telegraph line across Newfoundland. This project was much more important than may seem at first sight, for if it could be achieved it would reduce by several days the length of time it took for news to cross the Atlantic. Steamers from Europe would call at St. John's and any urgent messages could be flashed ahead of them along the telegraph to New York.

Unfortunately, building a line across Newfoundland is practically as difficult, because of the climate and the wild nature of the country, as laying a cable across the Atlantic. Even the original survey was bad enough; Gisborne reported 'my original party, consisting of six white men, were exchanged for four Indians; of the latter party, two deserted, one died a few days after my return, and the other has ever since proclaimed himself an ailing man'.

In the face of such hardships, it was not surprising that the Newfoundland Electric Telegraph Company went bankrupt in 1853 before more than forty miles of line had been erected. Gisborne, who had been left holding the company's debts, went to New York the next year in an attempt to raise more money for the scheme. By good fortune he met Cyrus Field, who was then relaxing after his South American trip and was not at first at all keen on becoming involved in any further business undertakings. He listened politely to Gisborne, but did not commit himself to any promise of help. Only the uncompleted line across Newfoundland was discussed, but when the meeting was over and he was alone in his library Field started to play with the globe and suddenly realised that the Newfoundland telegraph was merely one link in a far more important project. Why wait for steamers to bring news from Europe? Let the telegraph do the whole job. . . .

From that moment, Field became obsessed with the Atlantic telegraph. True, he was not the first man to conceive of a submarine cable linking Europe and America – we have already noted Morse's prediction – but he was the first to do anything practical. The next morning he wrote letters to Morse and to Lieutenant Maury, founder of the modern science of oceanography.

Matthew Fontaine Maury's classic book *The Physical Geography of the Sea* had not yet been published, but he was already famous – more famous, in the view of many of his superior officers, than a lieutenant should be. Though he had spent some years at sea, he had been accidentally lamed at the age of thirty-three and had then become head of the Depot of Charts and Instruments (the Hydrographic Office, as it is today). This had given him a unique opportunity of using his scientific talents, and by collating information from hundreds of ships' logs he had compiled the first detailed charts showing ocean currents and wind directions. These soon proved of immense value to navigators; by using Maury's charts, for example, ships rounding Cape Horn were able to cut their sailing time between New York and San Francisco from 180 days to a mere 133. Maury would have been astonished could he have known that, a century later, airline pilots were to profit in a similar way from a study of the winds, riding the jet streams of the stratosphere to span the continent in almost as many minutes as the old windjammers took days.

Unfortunately, the Lieutenant's services to his country and the world were not fully appreciated at higher levels; perhaps a series of scathing articles he had written about naval red tape had not helped his popularity. In 1855 a secret board, set up to conduct an efficiency and economy drive, put Maury on the permanent leave list. One would have thought that since his charts were now estimated to save several million dollars a year in reduced voyage times, the Navy could have afforded to keep him on its payroll.

A good many of Maury's contemporaries thought so too, and there was so much newspaper agitation that three years later the Navy was forced to reinstate him with the rank of Commander. There is an uncanny resemblance between the case of Lieutenant Maury and that of Admiral Rickhover, who dragged the United States Navy kicking and struggling into the Atomic Age and was duly passed over for promotion. But unfortunately for his future prospects, Maury – a Virginian – joined the losing side in the Civil War, and that was the end of his naval career.

By one of those coincidences which is inevitable when many people are thinking along the same lines, Maury received Field's letter at a moment when he had written to the Secretary of the Navy on the same subject. He had been forwarding a report of a recent survey of the North Atlantic, carried out by Lieutenant Berryman, which had disclosed the existence of a plateau between Newfoundland and Ireland. Maury had commented, in a letter to the Secretary of the Navy, February 22, 1854, that this plateau 'seems to have been placed there especially for the purpose of holding the wires of a submarine telegraph and of keeping them out of harm's way'.

Field could hardly have hoped for better news, and a few days later Morse called to see him with equally encouraging advice. With the world's greatest names in oceanography and telegraphy to back him up, Field now had only to convince the financiers.

Above: 1. HMS *Agamemnon* meets a whale when paying out the first
Atlantic cable, July 29, 1858.

Below: 2. The *Great Eastern* paying out cable, July 23, 1865.

Above: 3. The Prince of Wales watches the coiling of the cable in the *Great Eastern*'s tanks at Sheerness.

Below: 4. Lowering marker buoy after the cable had been lost, August 8, 1865.

This was not as difficult as it was to prove a few years later. Field's next-door neighbour in Gramercy Square, the influential millionaire Peter Cooper, gave him his support and this encouraged other capitalists to join in. The names of these far-sighted men are worth recording: besides Peter Cooper, they were Moses Taylor, Marshall O. Roberts and Chandler White.

With their backing, and the legal advice of his elder brother Dudley, Cyrus went to Newfoundland early in 1854 and took over the affairs of the moribund telegraph company. Its debts were paid, much local goodwill was thereby established, and Field obtained an exclusive charter for all cables touching Newfoundland and Labrador for the next fifty years. With this in his pocket, he returned triumphantly to New York, where it was the work of literally a few minutes (at six o'clock in the morning, which is not usually a good time to discuss business) for the subscribers to pledge $1,250,000 and float the 'New York, Newfoundland *and London* Telegraph Company'.

It took two and a half years of toil to substantiate the 'New York, Newfoundland' part of the company's title. Work was delayed for a whole year by the loss of the submarine cable that was to have spanned the St. Lawrence, but in 1856 the line was opened and the first part of Field's dream had come true. It was only a stepping-stone to his main objective, which had never been far from his mind.

One of his first acts had been to promote new surveys of the North Atlantic by both the British and American navies, which confirmed the existence of the so-called 'Telegraph Plateau'. It was not quite as smooth and flat as had been first supposed, but its changes of slope were no worse than those met with on many city streets. Submarine charts tend to be misleading in this respect, owing to the great exaggeration of the vertical scale. When one tries to show on one piece of paper a strip two thousand miles wide and five miles high, even the gentlest hills look like precipices. The encouraging thing about Telegraph Plateau, however, was not its relative flatness but the fact that its greatest distance from the surface was less than fifteen thousand feet – and submarine cables had already been laid in water as deep as this.

Field needed such encouragement; his troubles – financial and personal – were just beginning. While he was trying to raise

c

money for the projected cable, his only son died and at about the same time he lost his brother-in-law and business partner. It proved impossible to get the support he needed in the United States, which was now heading for one of its periodical depressions. So in 1856 Field sailed for England, in the hope that money would be less hard to find there.

At this point I cannot resist quoting from two modern books on telecommunications, leaving the reader to guess their countries of origin:

> The British capitalists were at first a bit timid about investing in what they considered an extravagant undertaking. . . .

> American big business had great difficulty in screwing up its courage to subscribe £27,000 and it was the merchants of Liverpool, Manchester, Glasgow, London and other British cities who only too willingly provided the rest. . . .

What actually happened was as follows. As soon as he arrived in England, Field arranged to meet the British telegraphic pioneers – notably John Brett, who was looking for fresh waters to conquer after his victory over the Channel. He also met the famous engineer Isambard Kingdom Brunel, then building the *Great Eastern*, which for half a century was to remain the mightiest ship that had ever moved upon the sea. In a prophetic moment, Brunel remarked to Field: '*There* is the ship to lay your cable.' Killed by his labours over the leviathan, the great engineer did not live to see his words come true ten years later, after both Field and the *Great Eastern* had survived endless disasters and defeats.

Fortunately for the project, Professor Morse was also in London at the time, and carried out a series of experiments which proved beyond doubt that signals could be sent through two thousand miles of cable. By connecting together ten circuits each of two hundred miles in length (the London to Birmingham line was used) Morse constructed a replica of the proposed Atlantic cable and succeeded in passing up to two hundred signals a minute through it.

This encouraging result convinced the British scientific world that the scheme was practical. Luckily, no one realised that the result was quite misleading; the line on which Morse conducted his tests was electrically much superior to the cable that was actually built and laid. It is not the first time that an over-optimistic report has launched a project and sustained its originators in the face of difficulties that they might never have faced had they known the facts.*

Armed with the evidence provided by his scientific experts, Field was now ready to tackle the British Government, as represented by the Navy and the Foreign Office. It is pleasant to record that he met neither scepticism nor, what is even more deadly, the 'all aid short of actual help' type of treatment. The Foreign Secretary, Lord Clarendon, was particularly interested in the project, but asked Field: 'Suppose you don't succeed? Suppose you make the attempt and fail – your cable is lost in the sea – then what will you do?' 'Charge it to profit and loss, and go to work to make another,' Field answered at once. It was an all too prophetic reply.

In the face of this optimism and perseverance, even the Treasury, that graveyard of lost hopes, gave its support. Within only a few days of explaining his scheme to its Secretary, Field received official promise of a Government subsidy of £14,000 a year, i.e. 4 per cent on the £350,000 capital which the project was expected to cost. The only condition was that the still-to-be-formed Atlantic Telegraph Company would carry any messages the British Government desired to send, giving them priority over all other traffic *except* that of the United States Government. The British Navy would also provide facilities for surveying the route and laying the cable.

The cast of characters for the forthcoming production was now assembled. The most important was a brilliant young telegraph engineer named Charles Tilston Bright, who at twenty-four now became chief engineer for one of the most ambitious projects of the century.

Charles Bright was another of those phenomenal Victorians who sometimes make one wonder if the human race has since deteriorated. When only nineteen, he had laid a complete system of telegraph wires under the streets of Manchester in a single

---

* Sometimes, of course, it works the other way – as when the German physicists decided, from incorrect measurements, that the atomic bomb was impossible.

night, without causing any disturbance to traffic. A year later, he had taken out twenty-four patents for basic inventions, some of which – such as the porcelain insulator for overhead wires – are still in use.

A man of action as well as a brilliant engineer, Bright became a Member of Parliament at thirty-three and died at the early age of fifty-five, burned out by his exertions. His monument is a network of telegraph cables stretching more than half-way round the globe and linking together all the countries of the world.

Bright had become interested in the Atlantic telegraph even earlier than Field. Between 1853 and 1855 he had conducted experiments to study the propagation of signals through two thousand miles of line, using for this purpose the ten circuits of two hundred miles each between London and Manchester, connected in series. In the summer of 1855 he had carried out a survey of the Irish coast and had decided that Valentia Bay, near the south-western tip of Ireland, was the best place to land a transatlantic cable. This decision has been endorsed by every company which has taken a cable to Ireland for the last hundred years.

A much less fortunate appointment was that of Dr. Edward Orange Wildman Whitehouse as the company's electrician. Dr. Whitehouse was a Brighton surgeon who had interested himself in telegraphy and had acquired a considerable knowledge of the subject by practical experimenting. He was a man of strong personality and fixed ideas, and although his enthusiasm did much to get the company started in its early days, his refusal to recognise his limitations was later to bring disaster.

The first meeting of the Atlantic Telegraph Company took place at Liverpool on November 12, 1856, and Field, Brett and Bright outlined the commercial prospects of the enterprise with such effect that the entire £350,000 was subscribed in a few days. Field took up £75,000 of this, not for his own benefit but on behalf – as he fondly imagined – of his fellow Americans. When he got back to his own country, however, he had the utmost difficulty in unloading even £27,000 of this amount, and was left holding the remainder himself.

Most of the capital was taken up by British business houses, though among the private subscribers it is interesting to note the names of Lady Byron and William Makepeace Thackeray. These literary figures were obviously keener on progress than their contemporary Thoreau, who had written in *Walden* two years before:

> We are in great haste to construct a magnetic telegraph from Maine to Texas; but Maine and Texas, it may be, have nothing important to communicate. We are eager to tunnel under the Atlantic and bring the Old World some weeks nearer to the New; but perchance the first news that will leak through into the broad, flapping American ear will be that Princess Adelaide has the whooping-cough. . . .

With the Atlantic Telegraph Company now organised, £350,000 in the bank, and the financial and material support of the British Government, Field returned to the United States at the end of 1856 fully confident that he would obtain equal support in his own country. However, when he approached President Buchanan to obtain the same terms that Britain had granted, he at once met violent opposition from Congress. As his brother Henry later remarked:

> He now found that it was much easier to deal with the English than with the American Government. . . . Those few weeks in Washington were worse than any icebergs off the coast of Newfoundland. The Atlantic cable has had many a kink since, but never did it seem to be entangled in such a hopeless twist as when it got among the politicians.

The arguments raised against what one would have thought to be a proposal of obvious and vital importance to the country now seem completely fantastic. (But let us not forget how later Congresses fought tooth and nail against the St. Lawrence Seaway for more than a quarter of a century.) Some senators objected to the enormous fee of $70,000 a year which the Government would have to pay for the privilege of swift and efficient transatlantic communication. Others thought that the State had no right to dabble in private business, and some objected to the proposed line because both ends were on British territory and the cable might therefore be put out of action in the event of war between the two countries. One Senator Jones of Tennessee opposed the scheme for the frank and forthright reason that 'he did not want anything to do with England or Englishmen'. There seemed a general fear (not yet wholly extinct in the United States) that if the British were keen on something there must be a catch in it, and any poor innocent Americans who became involved were liable to lose their shirts.

However, largely thanks to the support of Senator Thomas Rusk of Texas, the Bill was passed by a single vote on March 3, 1856. The United States Government granted the subsidy which would give the company a guaranteed source of income, and also made arrangements to provide ships to help with the cable-laying. A thankful but somewhat exhausted Cyrus Field hurried back to England to see how his British colleagues were faring.

They were making fine progress, spinning out cable at a rate which has seldom been matched since, and ought not to have been attempted then. Largely because Field had promised his backers that the telegraph would start working in 1857, specifications had been sent out to the manufacturers even before the board of management had been set up, and the production of the cable in the short time of six months was a remarkable performance. It involved drawing and spinning 335,000 miles of iron and copper wire and covering that with 300,000 miles of tarred hemp to form a cable 2,500 miles long. (The actual distance from Ireland to Newfoundland is about five hundred miles less than this, but the extra length was needed for slack in paying out, and to allow for possible losses.)

Progress, though swift, was far from smooth. Quite apart from the making of the cable, the expedition's ships had to be fitted out and a multitude of details supervised. Chief Engineer Bright, who was still twenty-four but ageing rapidly, commented in terms which will find an echo in the heart of anyone who has ever been engaged in what is so often rightly called a 'crash programme':

> At first one goes nearly mad with vexation at the delays; but one soon finds that they are the rule, and then it becomes necessary to feign a rage one does not feel. . . . I look upon it as the natural order of things that if I give an order it will not be carried out; or if by accident it is carried out, it will be carried out wrongly.

The company's engineers were not helped by streams of advice and criticism from outside experts, such as the Astronomer Royal, Sir George Airy, who stated dogmatically that 'it was a mathematical impossibility to submerge the cable successfully at so great a depth, and if it were possible, no signals could be transmitted through so great a length. . . .' When distinguished scientists made such fools of themselves, it is easy to excuse the numerous inventors who wrote to Bright with proposals based on the ancient fallacy that heavy objects did not sink to the sea-bed, but eventually came to rest at a level where their density was matched by that of the surrounding water. There is, of course, no truth in this idea, for water is so nearly incompressible that even at the greatest depths encountered in the ocean its density is only very slightly greater than at sea level.*

Some of the hopeful inventors wished to suspend the cable in mid-ocean by underwater parachutes or balloons; others even more optimistically wanted to connect it to a string of floating call-boxes across the Atlantic, so that ships could keep in touch with land as they crossed from continent to continent. Whether they were crazy or not, Charles Bright replied politely to all these proposals, few of which were inhibited by the slightest practical knowledge of the oceanographic and telegraphic facts of life.

The Atlantic Telegraph Company, in any event, had little need for outside help. On its own board of directors was a scientific genius (and for once the word is not misapplied) who was later to do more than any man to save the lost cause of submarine telegraphy and to retrieve the company's fortunes. Professor William Thomson had already come into collision with that opinionated amateur Dr. Whitehouse, and unluckily for all concerned his views had not prevailed over those of the project's official electrician.

* At the bottom of the Atlantic, the increase in density due to pressure alone is less than 2 per cent.

CHAPTER 5

Lord of Science

William Thomson, Lord Kelvin, was not the greatest scientist of the nineteenth century; on any reasonable list, he must come below Darwin and Maxwell, But it is probably true to say that he was the most famous man of science of his age, and the one whom the general public chiefly identified with the astonishing inventions and technical advances of the era.

In this, public opinion was correct, for Thomson was a unique bridge between the laboratory and the world of industry. He was an 'appled scientist' *par excellence*, using his wonderful insight to solve urgent practical problems. Yet he was very much more than this, being also one of the greatest of mathematical physicists. The range of his interests and activities was enormous; the multiplication of knowledge that has taken place since his time makes it impossible that we will see his like again. It would not be unfair to say that if one took half the talents of Einstein, and half the talents of Edison, and succeeded in fusing such incompatible gifts into a single person, the result would be rather like William Thomson. What his contemporaries thought of him is shown by the fact that he was the first scientist ever to be raised to the peerage.

Although we are really concerned with Thomson only as he affects the story of submarine telegraphy, he is such a fascinating and dynamic character that it is difficult to pass him by with no more than a glance. Moreover, it is impossible to understand his share of the story unless one has some appreciation of his extraordinary gifts and the use he made of them.

Heredity and environment between them left young William Thomson no chance of escaping from his destiny, even had he wished to do so. His father was Professor of Mathematics at Glasgow University, and from the tenderest age William was trained intensively for the academic life. He never went to school, all his teaching coming from his father, and as his mother died when he was only six years old the infant prodigy was clearly doomed to a life of interesting neuroses. In actual fact, about the only sign of Thomson's unorthodox upbringing was a certain lack of social graces and an inability to stop his brilliant mind from galloping off in all directions. No one can be certain that even these small defects can be blamed on his devoted father's training; moreover, as J. G. Crowther points out in his biography of Thomson, 'the indiscipline that hampered his scientific genius did not extend to his financial affairs'. It makes a most refreshing change to read of a scientist who accumulated a 128-ton yacht and a fortune of £162,000 by his own efforts. But, of course, Thomson was a Scot as well as a scientist.

Having matriculated at the age of ten, the young genius soon proved that his gifts were not limited to science. Two years later he won a prize for translating a dialogue of the Greek satirist Lucian – the author, incidentally, of the first interplanetary romance (*True History* – AD 160). And at the ripe age of sixteen, he produced a brilliant eighty-five-page essay, *On the Figure of the Earth*. For the benefit of any mathematically minded teenagers who feel like pursuing the matter further, this essay contained 'a discussion of the perturbation of the Moon's motion in longitude, and a deduction of the ellipticity from the constant of precession combined with Laplace's hypothetical law of density in the interior of the Earth'.

After such a start, it was not surprising that Thomson became Professor of Natural Philosophy at Glasgow University at the age of twenty-two. One of his earliest acts was to establish a physics laboratory in which the students could do practical work; this was the first such laboratory in Britain, if not in the world, and Thomson was granted the vast sum of £100 for the purchase of instruments.

As a lecturer, Thomson was not an unqualified success; indeed, when he had become Sir William, one of his students made the unkind comment: 'Behold the knight cometh, when no man can work.' However, by his personality and his genius he had an overwhelming effect upon two generations of physicists and engineers. Apart from his interest in telegraphy, he helped to lay the foundations of thermodynamics – the mathematical science of heat – and also conquered vast areas of magnetism, electricity and optics. His researches into astronomy and geophysics were also notable; some of his most famous calculations concerned the ages of the sun and earth, and he produced consternation among the geologists by claiming that the earth could not possibly be as old as they maintained; indeed, he set an upper limit of a miserable twenty million years to its age as a solid body.

This was one occasion when Thomson was completely wrong; he lived to witness the discovery of the energy source – radio-activity – which meant that the universe was much older than he had supposed. But by this time he was in his seventies and unable to appreciate the break-through into new fields of knowledge, so he never conceded his mistake.

Thomson became involved in the telegraph story as a result of his investigations into what are known as transient electric currents. What happens, he asked himself in 1853, when a battery is connected up to a circuit, in the minute interval of time before the current settles down to its steady value? At one moment nothing is happening; a fraction of a second later, a current of some definite amount is flowing. The problem was to discover what took place during the transition period, which is seldom as much as a hundredth of a second in duration, and is usually very much shorter.

Nothing could have seemed more academic and of less practical importance. Yet these studies led directly to the understanding of all electrical communication, and, some thirty years later, to the discovery of radio waves. If Thomson could have obtained a 5 per cent royalty on the use of the equations he derived, he would not have left a mere £162,000. He would have been the richest man on earth.*

Thomson showed that there were two possible ways in which a current could rise from zero to its steady value, depending upon the electrical characteristics of the circuit. A pendulum set swinging in a resisting medium – immersed in water, for example

* A title which was later applied, with some justice, to one of his pupils. Around 1885 Thomson met and befriended a brilliant young Armenian engineering student, who decided that he too would become a professor of physics. But the young man's father would not hear of this nonsense, and packed the boy off to the Middle East as soon as he had graduated. How different, one wonders, would modern history have been had Calouste Gulbenkian applied his extraordinary talents to science instead of using them to corner half the world's oil?

– gives a very exact analogy. If the friction is too great, the pendulum will drop slowly down to its point of rest without overshooting, but if the friction – the 'damping' – is sufficiently small, a whole series of oscillations of diminishing amplitude will take place. Precisely the same thing happens with the electric current, though this fact was not easy to demonstrate experimentally in the 1850s. Now we prove it in our homes a dozen times a day; when somebody switches on an electrical appliance and we hear the crackle on the radio, that is one of Thomson's oscillatory currents advertising its ephemeral presence.

A year later, using the same mathematical tools, Thomson began to investigate the behaviour of telegraph cables. It is possible to understand his main results, and to appreciate their importance, without any knowledge of the mathematics he used to obtain them. Putting it briefly, the problem involved was this: how long does it take for a signal to reach the far end of a telegraph cable?

It is a common error to imagine that electricity travels along a wire at the speed of light – 186,000 miles a second. This is never true, although in some circumstances this velocity can be approached. In most cases, the speed of a current is very much less than that of light – sometimes, indeed, only a tenth or a hundredth of its value.

This slowing-down is due to the electrical capacity of the line. There is no need to be alarmed by this phrase; it means exactly what it says. A telegraph cable behaves very much like a hosepipe; it takes a certain amount of electricity to 'fill it up' before there is any appreciable result at the other end.

Fortunately for the progress of the telegraphic art, this effect was of no practical importance in the early days of land lines. Their capacity was so low that messages passed through them without any appreciable delay, and it was not until the first submarine cables were laid across the English Channel and the North Sea that signal-retardation became a source of trouble. Its prime cause is the presence of the conducting sea-water which surrounds a cable and thus greatly increases its capacity. Because of this effect, a cable may need twenty times as much electricity to charge it up when it is submerged as it would require if suspended in air.

Thomson's analysis led him to his famous 'Law of Squares', which states that the speed with which messages can be sent through a given cable decreases with the square of its length. In other words, if one multiplies the length of a cable ten times, the rate of signalling will be reduced a hundredfold. This law is obviously of fundamental importance in long-distance submarine telegraphy; the only way of circumventing it is to increase the size of the conducting core.

This was not appreciated by all telegraph engineers, and was even denied by some – including, unfortunately Dr. Whitehouse. He had carried out experiments purporting to refute the law of squares; these had also led him to conclude that a small conducting wire might be better than a large one, which was the exact reverse of the truth.* When such confusion prevailed among the 'experts', it is hardly surprising that the first Atlantic cable was badly designed. It had about as much chance of success as a bridge built by engineers who did not understand the laws governing the strength of materials.

Thomson was only one of the company's directors, and had no authority – beyond his scientific prestige – over the men who were in charge of its technical affairs. He was thus in the difficult position, during the first act of the drama that was now beginning, of standing off-stage and making criticisms which the producer could ignore or accept as he saw fit.

Because of their determination to lay the cable during the summer of 1857, the promoters of the scheme had left no time for the experiments and tests which were essential for its success. The dynamic energy of Cyrus Field was partly responsible for this; when Thomson arrived on the scene, he discovered that the specifications for the cable had already been sent out to the manufacturers, and that it was now too late to alter them. What was more, when he had an opportunity of testing the completed article, he was shocked to discover that the quality of the copper varied so much that some sections conducted twice as well as others. There was nothing that could be done, except to insist that future lengths be made of the purest possible copper, and to hope that the existing cable would be good enough for the job.

* In Whitehouse's defence, it must be added that the same views had been expressed by Faraday and Morse.

The conductor itself consisted of seven strands of copper wire twisted together and insulated by three separate layers of gutta-percha (see Chapter 14). If there was a hole or imperfection in one layer, the other two would still provide adequate protection. Only in the extremely improbable event of three flaws occurring in exactly the same place would there be a danger of an electrical failure.

The insulated core was then covered with a layer of hemp, which in turn was armoured with eighteen strands of twisted iron wire. The resulting cable was about five-eighths of an inch in thickness, and weighed one ton per mile. This at once raised a serious problem, for the length needed to span the Atlantic weighed 2,500 tons – far too great a load to be carried in any single ship of the time.

The total cost of the cable was £224,000 – at least £1,000,000 by today's standards, though it is about as difficult to relate our present currency to the Victorian pound's real purchasing power as to that of the Russian rouble. The core was supplied by the Gutta Percha Company of Greenwich, London, but because of the time factor the armouring was divided between two firms: Glass, Elliott & Co., also of Greenwich, and Newall & Co., of Birkenhead. Owing to one of those slight oversights which can so easily wreck an enterprise of this kind, the wire armouring on the two halves of the cable was laid or twisted in opposite directions. This is a matter of great practical importance when it comes to joining together two cable-ends in mid-Atlantic; it is a little too late then to reverse one of the 1,250-mile-long sections so that all the heavy armouring corkscrews in the same direction at the point where you wish to make your splice.

The cable was completed within the remarkably short time of six months, and by July 1857 it was ready to go to sea. By rights, Whitehouse should have sailed with it, but at the last moment he pleaded ill health and Thomson was asked to fill the breach. It says much for the scientist's greatness of character that he agreed to do this, without any payment. The misshapen infant dumped on his doorstep was certainly not his baby, but he would give it the best start in life he could.

CHAPTER 6

False Start

To share the enormous weight of the cable between them, the warships *Niagara* and *Agamemnon* had been provided by the United States and British Governments respectively. The ninety-one-gun *Niagara* was the finest ship in the American Navy; the largest steam frigate in the world, she had lines like a yacht and her single screw could drive her with ease at twelve miles an hour. The *Agamemnon*, on the other hand, would not have looked out of place at Trafalgar; she was one of the last of the wooden walls of England, and though she had steam power as well as sail one would not have guessed it by looking at her.

Both ships had been extensively modified to allow them to carry and pay out their 1,250 tons of cable. Their holds had been enlarged into circular wells or tanks in which the cable could be coiled; even so the *Agamemnon* was forced to carry several hundred tons of it on deck – a fact which later brought her to the edge of disaster.

The *Niagara* was carrying two Russian officers as observers, which could not have been too pleasing to the British, since the Crimean War had finished only a year before. There were no journalists on board, this being against service regulations. Perhaps the American Navy was still smarting under the impact of a recent devastating exposé in a book named *White Jacket*, though it could take some consolation from the knowledge that the author's latest work, a tedious novel called *Moby Dick*, had been a complete flop.

The British had no such inhibitions, and their share of the enterprise was fully covered by the Press. It is from an issue of the London *Times* dated July 24, 1857, that we have salvaged this account of a banquet given to the workmen of the cable company and the crew of the *Agamemnon* just before she was due to sail; its subtle nuances say more about the manners of the time than many volumes of social history:

The manufacturers provided a magnificent banquet for the guests, and a substantial one for the sailors . . . by an admirable arrangement, the guests were accommodated at a vast semi-circular table, while the sailors and workmen sat at a number of long tables arranged at right-angles with the chord, so that the general effect was that all dined together, while at the same time sufficient distinction was preserved to satisfy the most fastidious. . . .

After loading their respective halves of the cable the two warships (with their escorts the *Susquehanna* and *Leopard*) sailed to their rendezvous at Valentia Bay, County Kerry. The plan that had been adopted, at the insistence of the directors, was for the *Niagara* to lay the whole of her cable westward from Ireland, and for the *Agamemnon* to splice on in mid-Atlantic and then complete the job. This would have the advantage that the expedition would be in continual contact with land and could report progress through the unwinding cable all the way across the Atlantic. On the other hand, if the ships arrived in mid-ocean during bad weather, and it was impossible to make the splice, half the cable would be lost.

The arrival of the telegraph fleet in this remote corner of Ireland attracted crowds of sightseers, and the local nobility rose to the occasion with inspiring speeches. Everyone realised the importance of the event, and when the shore end of the cable was landed on August 5, 1857, Henry Field reports:

Valentia Bay was studded with innumerable small craft, decked with the gayest bunting – small boats flitted hither and thither, their occupants cheering enthusiastically as the work successfully progressed. The cable boats were managed by the sailors of the *Niagara* and *Susquehanna*, and it was a well-designed compliment, and indicative of the future fraternisation of nations, that the shore-rope was arranged to be presented at this side of the Atlantic to the representative of the Queen, by the officers and men of the United States Navy, and at the other side British officers and sailors would make a similar presentation to the President of the great Republic.

For several hours the Lord Lieutenant of Ireland stood on the beach, surrounded by his staff and the directors of the cable company, watching the arrival of the cable, and when at length the American sailors jumped through the surge with the hawser to which it was attached, his Excellency was among the first to lay hold of it and pull it lustily to shore. . . .

It was too late to set sail that day, so the cable-laying began the next morning – Thursday, August 6, 1857. Almost at once there was a minor but annoying setback; five miles out, the cable caught in the primitive paying-out mechanism and broke. It was necessary to go back to the beginning, lift the section that had already been laid, and run along it until the break was reached.

At length [continues Henry Field], the end was lifted out of the water and spliced to the gigantic coil [i.e. the 1,250 miles in the *Niagara's* hold] and as it dipped safely to the bottom of the sea, the mighty ship began to stir. At first she moved very slowly, not more than two miles an hour, to avoid the danger of accident; but the feeling that they are away at last is itself a relief. The ships are all in sight, and so near that they can hear each other's bells. The *Niagara*, as if knowing that she is bound for the land out of whose forests she came, bends her head to the waves, as her prow is turned towards her native shores.

Slow passed the hours of that day. But all went well, and the ships were moving out into the broad Atlantic. At length the sun went down in the west and stars came out on the face of the deep. But no man slept. A thousand eyes were watching a great experiment as those who have a personal interest in the issue. . . . There was a strange, unnatural silence in the ship. Men paced the deck with soft and muffled tread, speaking only in whispers, as if a loud voice or a heavy footfall might snap the vital chord. So much have they grown to feel for the enterprise, that the cable seemed to them like a human creature, on whose fate they hung, as if it were to decide their own destiny. . . .

So it went for the next three days. Let us, without any perceptible change of style, switch narrators in mid-ocean from the American Henry Field to the quintessentially British London *Times*:

The cable was paid out at a speed a little faster than the ship, to allow for any inequalities on the bottom of the sea. While it

Above: 5. Machines covering the Atlantic cable wire with gutta percha at the Gutta Percha Company's works, Wharf-road, London.

Below: 6. The method of taking the Atlantic Telegraph on board – sketched from the stern gallery of HMS *Agamemnon*.

Above: 7. Landing the Crimean cable, 1855. Electric telegraph from the camp before Sebastopol to London. Landing and connecting the wire cable at Cape Kaliakra.

Below: 8. The paying-out machinery on the *Great Eastern*.

was thus going overboard, communication was kept up constantly
with the land. Every moment the current was passing between
ship and shore. . . . On Monday they were over 200 miles to
sea. They had got far beyond the shallow waters of the coast.
They had passed over the submarine mountain . . . where Mr.
Bright's log gives a descent from 550 to 1,750 fathoms within
eight miles.* Then they came to the deeper waters of the
Atlantic, where the cable sank to the awful depth of two
thousand fathoms. Still the iron cord buried itself in the wave,
and every instant the flash of light in the darkened telegraph
room told of the passage of the electric current. . . .

But not for much longer – for at 9 a.m. that morning the line
suddenly went dead. There was a gloomy consultation among the
engineers, and all hope had been abandoned when, quite unexpectedly, signals started coming through again. This two-and-
a-half-hour break in continuity was never satisfactorily explained;
it might have been due to a faulty connection in the equipment at
either end, or to a flaw in the cable itself.

This was a disturbing setback, but the next day brought
catastrophe. The cable had been running out so rapidly (at six
miles an hour against the ship's four) that it was necessary to
tighten the brake on the paying-out mechanism. By an unfortunate error, the tension was applied too suddenly, and the cable
snapped under the strain.

There was nothing to do but to postpone the attempt until the
next year, since the amount of cable in the tanks was not sufficient
to risk another try. But Field and his colleagues, though disappointed, were not despondent. They had successfully laid 335
miles of cable, a third of it in water more than two miles deep, and
had been in telegraphic communication with land until the
moment the line had parted. This proved, it seemed to them,
that there was nothing impossible in the job they were attempting.

The ships returned to England and unloaded the 2,200 miles of
cable they had brought back. It was coiled up on a wharf at
Plymouth to await the next expedition, while the *Niagara* and

* This figure sounds very impressive – but a gradient of one in five is
hardly precipitous. There is no need to go to San Francisco to find
city streets that are steeper than this.

D

Agamemnon reverted to their naval duties – somewhat handi-capped by the gaping holes that had been torn in their entrails.

Meanwhile, the mistakes revealed by the first attempt were studied by the engineers to prevent them occurring again. The paying-out mechanism, which had been the prime cause of the failure, was completely redesigned. A new type of friction brake was used which would automatically release if too much tension was applied; we read with morbid fascination that 'this clever appliance had been introduced in connection with the crank apparatus in gaols, so as to regulate the amount of labour in proportion to the strength of the prisoner'.

The indefatigable Field had returned to America to raise more money, only to find a depression sweeping the country and much of his fortune lost. The failure of the first expedition had also shaken confidence in the project and it was hard to obtain support on either side of the Atlantic; nevertheless the new capital was secured and seven hundred miles of fresh cable was ordered.

While the preparations were going ahead for the next expedition, Professor Thomson was far from idle. In addition to his normal work at the university, he continued to study the problem of the Atlantic telegraph. One important fact that had emerged from his mathematical analysis was that if a sufficiently sensitive detector could be used at the receiving end of the cable, the rate of signalling would be much increased. This results from the fact that when a sudden electrical impulse (say a dot or a dash) is applied to one end of a cable, it does not appear at the other end as an equally abrupt rise of voltage. The first intimation at the receiver is a gently rising wave of electricity which takes an appreciable time to reach its maximum value. If the first onset of this wave could be sensed by a sufficiently delicate instrument, there would be no need to wait for the crest of the wave to arrive. The signal would have been detected, and the next one could now be sent.

The way in which the clearly defined pulse of electricity from, say, a Morse key is 'smeared out' as it progresses along a submarine cable can be appreciated from this analogy. Think of the water behind a dam; it forms a vertical wall, which we can liken to the original sharp-edged pulse sent into the cable. The moment of transmission corresponds to the sudden breaking of the dam; at once the water starts to collapse – to flatten out. At a point a considerable distance away, the first intimation that anything has happened is an almost inconspicuous wave that may take a considerable time to build up to its maximum value. And once you have noticed this little wave, there is no point in waiting any longer. You know what's coming.

Thomson's objective, therefore, was extreme detector sensitivity. But Whitehouse, with his remarkable talent for doing the wrong thing, took just the opposite approach. He proposed to use brute force at the transmitting end of the cable, pushing so much current into it that even insensitive instruments – such as his own patent automatic printer – could read the messages sent. The result of this policy we will see in due course.

The solution to the receiving problem was given by Thomson's monocle; he was twirling it one day, and noticed how swiftly the reflected light danced around the room. This led him to his famous mirror galvanometer, in which the minute deflection of a coil carrying an electric current was greatly magnified by a spot of light reflected from a tiny mirror attached to it. Thomson had, in effect, invented an instrument with a weightless pointer.

In passing, it might be mentioned that the story of Thomson's monocle appears more authentic than that of Newton and the apple (though there are grounds for believing that that also may be true). And discoveries prompted by such chance observations are never accidents; they only happen to those who have been thinking long and earnestly about a problem, and whose minds are therefore in a sensitive, receptive mood. How many philosophers before Newton had seen apples falling? How many bacteriologists before Fleming had noticed inexplicable moulds on their cultures?

The mirror galvanometer, because of its delicacy, simplicity and elegance made a great impression on Thomson's contemporaries. Clerk Maxwell, one of whose relaxations was writing light verse, produced a Tennysonian parody in celebration which opens:

> The lamplight falls on blackened walls
> And streams through narrow perforations
> The long beam trails o'er pasteboard scales
> > With slow, decaying oscillations.
> Flow, current, flow! Set the quick light-spot flying!

> Flow, current, answer light-spot, flashing, quivering,
>   dying. . . .

In the spring of 1858, the great enterprise got under way again. Once more the *Agamemnon* and the *Niagara* were commissioned as cable-layers, and the Admiralty provided the sloop *Gorgon* as an escort. The United States Navy had promised that the *Susquehanna* would also be available as in the previous year, but she was quarantined in the West Indies with yellow fever aboard.

As soon as he heard this news – which threatened the success of the whole project – Field promptly buttonholed the First Lord of the Admiralty, and apologetically asked if the British Navy could add a third vessel to the two it was already providing. The First Lord explained that the Navy was so short of ships that it was hiring them, but promised to do his best. Within a few hours the *Valorous* had been made available; the Victorians could move quickly when they wanted to, even without benefit of telephones.

This time, at the insistence of the engineers, it was decided to start from mid-Atlantic and let the ships lay the cable in opposite directions. Not only would this be more economical in time, but it would mean that the all-important splice could be made at leisure, when weather conditions were most suitable.

After some initial tests in the Bay of Biscay (where, almost a hundred years later, the components of the Atlantic telephone cable also had their baptism of deep water) the little fleet sailed from Plymouth under fair skies on June 10, 1858. Once again Whitehouse had asked to be excused on medical grounds, and once again Thomson took his place (unpaid). It was lucky for Whitehouse that he stayed on land, for only two days after they had left harbour beneath clear skies, the four ships ran into one of the worst Atlantic storms ever recorded.

They were scattered over the face of the sea, each ship fighting desperately for its life. The *Agamemnon* was in particular danger, being made almost unmanageable owing to the 1,300 tons of cable in her hold and, a still more serious hazard, the 250 tons coiled on deck. Thanks to Nicholas Woods, correspondent of the London *Times*, we have an account of the storm which must be among the most vivid in the literature of the sea. Listen to his description of the *Agamemnon* in her hour of peril:

The massive beams under her upper deck coil cracked and snapped with a noise resembling that of small artillery, almost drowning the hideous roar of the wind as it moaned and howled through the rigging. . . . At 4 a.m. sail was shortened – a long and tedious job, for the wind so roared and howled, and the hiss of the boiling sea was so deafening, that words of command were useless, and the men aloft, holding on with all their might to the yards as the ship rolled over and over almost to the water, were quite incapable of struggling with the masses of wet canvas, that flapped and plunged as if men and yards and everything were going away together. . . . At about half-past ten o'clock three or four gigantic waves were seen approaching the ship, coming slowly on through the mist nearer and nearer, rolling on like hills of green water, with a crown of foam that seemed to double their height. The *Agamemnon* rose heavily to the first, and then went lower quickly into the deep trough of the sea, falling over as she did so, so as almost to capsize completely. There was a fearful crashing as she lay over this way, for everything broke adrift . . . a confused mass of sailors, boys, marines, with deck-buckets, ropes, ladders and everything that could get loose, were being hurled in a mass across the ship. . . . *The lurch of the ship was calculated at forty-five degrees each way for five times in rapid succession.* . . . The coil in the main hold . . . had begun to get adrift, and the top kept working and shifting as the ship lurched, until some forty or fifty miles were in a hopeless state of tangle, resembling nothing so much as a cargo of live eels. . . .

The sun set upon as wild and wicked a night as ever taxed the courage and coolness of a sailor. . . . The night was thick and very dark, the low black clouds almost hemming the vessel in; now and then a fiercer blast than usual drove the great masses slowly aside, and showed the moon, a dim, greasy blotch upon the sky, with the ocean, white as driven snow, boiling and seething like a cauldron. But these were only glimpses, which were soon lost, and again it was all darkness, through which the waves, suddenly upheaving, rushed upon the ship as though they must overwhelm it. . . . The grandeur of the scene was almost lost in its dangers and terrors, for of all the many forms in which death approaches man there is none so easy in fact, so terrific in appearance, as death by ship-wreck. . . .

But all things have an end, and this long gale – of over a week's duration – at last blew itself out, and the weary ocean rocked itself to rest. . . . As we approached the place of meeting the angry sea went down. The *Valorous* hove in sight at noon; in the afternoon the *Niagara* came in from the north; and at even, the *Gorgon* from the south; and then, almost for the first time since starting, the squadron was reunited near the spot where the great work was to have commenced fifteen days previously – as tranquil in the middle of the Atlantic as if in Plymouth Sound.

After this ordeal, one would have thought that the expedition had earned the right to success. The battered vessels were made shipshape, the cable-ends were spliced together, and on June 26 the *Niagara* sailed west for Newfoundland and the *Agamemnon* headed east towards Ireland.

They had gone only three miles when the cable fouled the paying-out machinery on board the *Niagara* and snapped. This was anti-climax number one, but nobody was too upset as little time and cable had been lost.

On the second attempt, the ships got eighty miles apart before anything went wrong. Then they suddenly lost telegraphic contact, and each assumed that the cable had broken aboard the other. They hurried back to the rendezvous and hailed each other simultaneously with the words: 'How did the cable part?' It was very disconcerting to find no explanation for what had happened; for some unknown reason, the cable had broken on the sea-bed.

A third time the splice was made, and, no doubt with all aboard wondering when they would meet again, the ships sailed apart once more. Unfortunately, it was not a case of third time lucky. After two hundred miles had been paid out, the cable parted on the *Agamemnon*. The ships were now short of provisions, and according to prearranged plans they headed back independently to Ireland for a council of war.

It was an unhappy board of directors which met to consider the next move. Some, in despair, wished to sell the remaining cable and abandon the whole enterprise. But Field and Thomson argued for a fresh attempt, and in the end their counsel prevailed. The faint-hearted directors resigned in disgust at such stubborn foolishness, and by July 29 the ships were back in mid-Atlantic, ready for the fourth try.

There was no ceremony or enthusiasm this time when the splice went overboard and the ships parted. Many felt that they were on a fool's errand; as Field's brother remarked in his memoirs: 'All hoped for success, no one dared to expect it.'

And certainly no one could have guessed that they were about to achieve, in the highest degree, both success and failure.

CHAPTER 7

Triumph and Disaster

It was just as well for the American Press that it had no representative on board the *Niagara*, for the westward voyage was a
monotonously peaceful one with the cable paying out uneventfully
hour after hour. The only excitement was in the electricians'
cabin, for twice during the week the signals from the *Agamemnon*
failed but came back again in full strength after a few hours'
anxiety. Apart from this, the *Niagara's* log records 'light breeze
and moderate sea' almost all the way, until the moment she
arrived in Trinity Bay, Newfoundland, with her 1,300 miles of
cable safely strung across the bed of the Atlantic.

The eastward-sailing *Agamemnon*, on the other hand, had once
again had an adventurous voyage, and several times had skirted
mechanical or electrical disaster. Considering the conditions
under which Thomson and his assistants worked, it is astonishing
that they were able to keep their instruments operating at all.
Listen to this description of the telegraph room as given by the
Sydney Morning Herald:

> The electrical room is on the starboard side of the main deck
> forward. The arrangements have been altered several times in
> order to avoid the water which showers down from the upper
> deck. At one end of the little place the batteries are ranged on
> shelves and railed in. . . . The most valuable observation is
> taken in sending on the marine galvanometer. Three seconds
> before it is taken, the clerk who times all the observations by a
> watch regulated by a chronometer too valuable to be brought
> into so wet a place says 'Look out'. The other clerk at once
> fixes his eye on the spot of light, and immediately the word is
> given 'Now' records the indication. This testing is made from
> minute to minute, so that a flaw is detected the moment it
> occurs.

The ships had spliced the cable on July 29, 1858, midway between Europe and America, in water 1,500 fathoms deep. To let *The Times* continue the story:

> For the first three hours the ships proceeded very slowly, paying out a great quantity of slack, but after the expiration of this time, the speed of the *Agamemnon* was increased to about five knots, the cable going at about six. . . . Shortly after six o'clock a very large whale was seen approaching the starboard bow at a great speed, rolling and tossing the sea into foam all around . . . it appeared as if it were making direct for the cable, and great was the relief of all when the ponderous living mass was seen slowly to pass astern, just grazing the cable where it entered the water. . . .

A few hours later, there was a real crisis, vividly depicted by the *Sydney Morning Herald*'s reporter:

> We had signalled the *Niagara* '40 miles submerged' and she was just beginning her acknowledgment when suddenly, at 10 p.m., communication ceased. According to orders, those on duty sent at once for Dr. Thomson. He came in a fearful state of excitement. The very thought of disaster seemed to over-power him. His hand shook so much that he could scarcely adjust his eyeglass. The veins on his forehead were swollen. His face was deathly pale. After consulting his marine galvanometer, he said the conducting wire was broken, but still insulated from the water. . . . There did not seem to be any room for hope; but still it was determined to keep the cable going out, that all opportunity might be given for resuscitation. The scene in and about the electrical room was such as I shall never forget. The two clerks on duty, watching, with the common anxiety depicted on their faces, for a propitious signal; Dr. Thomson, in a perfect fever of nervous excitement, shaking like an aspen leaf, yet in mind keen and collected, testing and waiting. . . . Mr. Bright, standing like a boy caught in a fault, his lips and cheeks smeared with tar, biting his nails and looking to the Professor for advice. . . . The eyes of all were directed on the instruments, watching for the slightest quiver indicative of life. Such a scene was never witnessed save by the bedside of the dying. . . .

Dr. Thomson and the others left the room, convinced that they were once more doomed to disappointment. . . .

But they were not. No one ever knew exactly what had happened; perhaps the cable's conducting core had broken under the strain of laying, but reunited on the sea-bed when the tension was relaxed, and the elasticity of the coverings brought the wires together again. In any event, the signals returned at last, and the cable spoke again.

Our joy was so deep and earnest that it did not suffer us to speak for some seconds. But when the first stun of surprise and pleasure passed, each one began trying to express his feelings in some way more or less energetic. Dr. Thomson laughed right loud and heartily. Never was more anxiety compressed into such a space. It lasted exactly one hour and a half, but it did not seem to us a third of that time. . . .

The ship now began to run into heavy seas, and started to pitch and roll in a manner that put a great strain on the cable.

During Sunday the sea and wind increased, and before the evening it blew a smart gale. Now indeed were the energy and activity of all engaged in the operation tasked to the utmost . . . the engineers durst not let their attention be removed from their occupation for one moment, for on their releasing the brake on the paying-out gear every time the stern of the ship fell into the trough of the sea entirely depended the safety of the cable. . . . Throughout the night, there were few who had the least expectation of the cable holding on till morning, and many remained awake, listening for the sound that all most dreaded to hear – namely, the gun which should announce the failure of all our hopes. But still the cable, which, in comparison with the ship from which it was paid out, and the gigantic waves among which it was delivered, was but a mere thread, continued to hold on, only leaving a silvery phosphorus line upon the stupendous seas as they rolled on towards the ship. . . .

Quite apart from the extreme danger to the cable, the need to maintain speed caused the supply of coal to dwindle at an alarming rate. At one time it looked as if it would be necessary to start burning up the spars and planking in a *grand finale* like the last lap of *Around the World in Eighty Days*. But luckily the gale slowly abated; both the *Agamemnon* and her cable had weathered the storm.

There was a brief flurry of excitement towards the end of the voyage when an inquisitive American barque bore down upon the telegraph fleet as it ploughed along on its predetermined and unalterable course. The escorting *Valorous* had to fire her guns to scare away the interloper, who was doubtless surprised by such a rude reception. Luckily, no international incident resulted from this display of arms, though as *The Times* put it: 'Whether those on board her considered that we were engaged in some filibustering expedition, or regarded our proceedings as another British outrage against the American flag, it was impossible to say; but in great trepidation she remained hove-to until we lost sight of her.'

But at last, on the morning of Tuesday August 5:

the bold and rocky mountains which entirely surround the wild and picturesque neighbourhood of Valentia, rose right before us at a few miles distance. Never, probably, was the sight of land more welcome, as it brought to a successful termination one of the greatest, but at the same time, most difficult, schemes which was ever undertaken. Had it been the dullest and most melancholy swamp on the face of the earth that lay before us, we would have found it a pleasant prospect; but as the sun rose from the estuary of Dingle Bay, tingling with a deep soft purple the lofty summits of the steep mountains which surround its shores, and illuminating the masses of morning vapour which hung upon them, it was a scene which might vie in beauty with anything that could be produced by the most florid imagination of an artist.

No one on shore was apparently conscious of our approach, so the *Valorous* steamed ahead to the mouth of the harbour and fired a gun. . . . As soon as the inhabitants became aware of our approach, there was a general desertion of the place, and hundreds of boats crowded around us, their passengers in the greatest state of excitement to hear all about our voyage. . . . Soon after our arrival, a signal was received from the *Niagara* that they were preparing to land, having paid out one thousand and thirty nautical miles of cable, while the *Agamemnon* had accomplished her portion of the distance with an expenditure of one thousand and twenty miles, making the total length of the wire submerged two thousand and fifty geographical miles. . . .*

Dr. Thomson came into the electrical cabin, evidently in a state of enjoyment so intense as almost to absorb the whole soul and create absence of mind. His countenance beamed with placid satisfaction. He did not speak for a little, but enjoyed himself stretching scraps of sheet gutta-percha over the hot globe of our lamp, watching them with an absent eye as they curled and shrank. . . . When we got close inshore we threw off the cable-boat. Before her prow grated on the strand her impetus had taken her ashore. The *Valorous*, in the distance, fired her guns. The end was seized by the jolly tars and run off with; a good humoured scuffle ensued between them and the gentlemen of the island for the honour of pulling the cable up into the office. The Knight of Kerry was upset in the water. . . .

Europe and America had at last been linked together. The news of this completely unexpected success, when everyone except a few enthusiasts had been convinced that the enterprise was hopeless, created a sensation. To read the papers of the time, one would think that the millennium had arrived. Even the staid *Times*, not prone to hyperbole, informed its readers: 'The Atlantic is dried up, and we become in reality as well as in wish one country. . . . The Atlantic Telegraph has half undone the Declaration of 1776, and has gone far to make us once again, in spite of ourselves, one people. . . .'

There were, of course, celebrations all over the United States; countless sermons were preached, many of them based on the Psalmist's verse: 'Their line is gone out through all the earth, and their words to the end of the world.' Inspiring poetry, almost as good as that which can now be produced by any self-respecting electronic computer, was churned out by the yard to fit the

---

* This is an error; the reporter had forgotten that the nautical mile is 15 per cent longer than the geographical mile, so that the total length of cable laid was about 2,350 miles. The actual great-circle distance between the two ends of the cable was 1,950 miles, the difference being due to the slack or excess cable which had to be laid to follow the contours of the sea-bed.

occasion. Enthusiasm was sustained despite the long delay in opening the service to the eagerly waiting public; this delay, it was explained, was due to the delicacy of the instruments and the careful adjustment they required.

When a message from Queen Victoria to President Buchanan was received on August 16 further rejoicings and demonstrations broke out, to such effect that the roof of the New York City Hall was ignited by the fireworks and the whole structure was barely saved from the flames. In England, Charles Bright received a knighthood at the early age of twenty-six for his work as chief engineer of the project; in New York, on September I, Cyrus Field was given a vast public ovation – at the very moment, ironically enough, when the Atlantic Telegraph had given up the ghost.

For the cable which had been laid with such expense and difficulty, and after so many failures, was slowly dying. Indeed, when one considers the imperfections in its manufacture, and the various ordeals it had gone through, it is astonishing that it had ever worked at all.

In his effort to prove that no direct Atlantic line could be an economic proposition, Colonel Tal Shaffner was later to produce a full transcript of the 1858 cable's working. It is a record of defeat and frustration – a four-week history of fading hopes. Even after five days had been allowed for setting up the receiving and transmitting equipment, this log of *all* the messages sent from Newfoundland to Ireland on the whole of the sixth day speaks for itself:

'Repeat, please'
'Please send slower for the present.'
'How?'
'How do you receive?'
'Send slower.'
'Please send slower.'
'How do you receive?'
'Please say if you can read this.'
'Can you read this?'
'Yes.'
'How are signals?'

'Do you receive?'
'Please send something.'
'Please send V's and B's.'
'How are signals?'

This, remember, is an entire day's working. It was not until more than a week after the laying of the cable-ends that the first complete message got through. Part of the delay was due to the fact that as soon as intelligible signals were received from Newfoundland on Thomson's mirror galvanometer, Whitehouse at the Valentia end immediately had his patent automatic recorder switched into the circuit. This instrument functioned adequately on short lines, but was quite incapable of interpreting the minute and distorted signals which were trickling through the injured cable.

There was similar confusion over the sending of the signals. Whereas Thomson wished to use low-voltage batteries to provide power for signalling, Whitehouse insisted on employing the huge induction or spark coils he had built, which were five feet long and developed at least two thousand volts. The use of these coils was to result in a great deal of public controversy when the cable finally failed, and there can be little doubt that they helped to break down the faulty insulation.

It was nine days before a single word got through the cable from east to west, but on the twelfth day (August 16) the line was working well enough to start transmitting a ninety-nine-word message of greetings from Queen Victoria to President Buchanan. It took sixteen and a half hours before the message was completed; it would arrive in America nearly as quickly today by airmail.

The first commercial message ever telegraphed across the Atlantic was sent the next day (August 17) from Newfoundland to Ireland. It is one which we can still fully appreciate: 'Mr. Cunard wishes telegraph McIver Europa collision Arabia. Put into St. John's. No lives lost.'

More days went by while the operators struggled to keep in contact and to transmit the messages which were piling up at either end. Sometimes a personal note intruded, as when Newfoundland remarked plaintively to Ireland, 'Mosquitoes keep biting. This is a funny place to live in – fearfully swampy' or when

Thomson, no doubt after turning the Valentia office upside down, was forced to ask Newfoundland: 'Where are the keys of the glass cases and drawers in the apparatus room?' (The helpful answer: 'Don't recollect.')

Finally, after Newfoundland had signalled, 'Pray give some news for New York, they are mad for news,' the first Press dispatch was successfully sent on the twenty-third day (August 27). It is interesting to compare the headlines of 1858 with those of a hundred years later: 'Emperor of France returned to Paris Saturday. King of Prussia too ill to visit Queen Victoria. Her Majesty returns to England August 31. Settlement of Chinese question. Chinese empire open to trade; Christian religion allowed; foreign diplomatic agents admitted; indemnity to England and France. Gwalior insurgent army broken up. All India becoming tranquil.'

Yes, a lot has happened since that far-off summer. Where are the Emperor of France and the King of Prussia today? And if there had been a United Nations back in 1858, it would have been an indemnity *from* England and France.

The final reference in the message is to the Indian Mutiny, which was then nearing its end. It was in connection with the Mutiny that the cable gave a dramatic proof of its value; only a day before it broke down completely it carried orders countermanding the sailing of the 62nd Regiment from Nova Scotia to India, where it was no longer needed. This single message was estimated to have saved the War Office no less than £50,000 – one-seventh of the entire cost of the cable.

The last message passed through the cable at 1.30 p.m. on September 1; it was, ironically enough, a message to Cyrus Field at the banquet in his honour in New York, asking him to inform the American Government that the company was now in a position to forward its messages to England. . . .

Thereafter, all was silence. After their brief union, the continents were once more as far apart as ever. The Atlantic had swallowed up the months of toil, the 2,500 tons of cable, the £350,000 of hardly raised capital.

The public reaction was violent, and those who had been most fervent in their praise now seemed ashamed of their earlier enthusiasm. Indeed, it was even suggested that the whole affair had been a fraud of some kind – perhaps a stock manipulation on the part of Cyrus Field. One Boston newspaper asked in a trenchant headline, 'Was it a hoax?' and an English writer proved that the cable had never been laid at all.

What had been hailed as the greatest achievement of the century had collapsed in ruins; it was to be eight long years before Europe and America would speak to each other again across the bed of the ocean.

CHAPTER 8

Post-mortem

With the failure of the 1858 cable, a third of a million pounds of private capital had been irretrievably sunk into the Atlantic. Within a year, the submarine telegraph engineers had achieved an even more resounding catastrophe. A cable through the Red Sea to India, laid at a cost of £800,000, had also failed completely – and this time it was Government money that was lost. It is not surprising that there was a general public outcry, much correspondence in the London *Times*, and a demand for a commission of inquiry.

The report of this commission, which sat from December 1859 to September 1860, must be one of the more monumental publications of HM Stationery Office. Running to over five hundred foolscap pages of small print, it is longer than the Bible. Its title is correspondingly impressive: 'Report of the Joint Committee appointed by the Lords of the Committee of Privy Council for Trade and the Atlantic Telegraph Company to inquire into the Construction of Submarine Telegraph Cables: together with the Minutes of Evidence and Appendix.'

There is a strange resemblance, across almost a century of time, between the inquiry into the failure of the Atlantic cable and that which took place in 1954 to discover the causes of the disasters to the Comet jet airliners. In each case British engineering prestige was at stake, immense sums had been lost, great hopes had been raised and then dashed to the ground. And it would not be unfair to say that the final verdicts were similar: the daring of the engineers had outrun their knowledge of a new technology.

The report of the Privy Council, which was issued in April 1861, is not only a summary of the electrical art a hundred years ago; it provides fascinating glimpses of the personalities involved, from the mighty Professor Thomson to the unfortunate Dr. Whitehouse, who was widely blamed as chief architect of the disaster. It also contains many proposals for designing or laying submarine

E

cables which are entertainingly absurd, and not a few which were more prophetic than their authors could ever guess.

For example, one Captain Selwyn, RN, wanted to avoid paying out cable from tanks inside the ship (with the attendant risk of kinks and breakages) by having it wound on a large floating drum which would be towed behind a steamer. The drum would revolve in the water as the cable uncoiled, but the committee remarked: 'We have great doubts as to the practicability of this plan.'

As far as the open Atlantic was concerned, the committee was quite correct. However, in 1944 just such floating drums were used to lay the underwater pipeline PLUTO (*P*ipe *L*ine *U*nder *T*he *O*cean) through which fuel could be pumped across the English Channel to power the invasion of Europe.

The committee consisted of eight members, four appointed by the Board of Trade and four appointed by the Atlantic Telegraph Company, which thus had one foot in the dock and the other on the judges' bench. But though there was occasional acrimony in the evidence given, there seems to have been no whitewashing; the 1861 report arrived at the facts, and its masses of technical information marked the transition of submarine telegraphy from guess-work to science.

Of the eight men who spent almost a year listening to half a million words of evidence, only one is remembered today. He is Professor Charles Wheatstone, whose contributions to telegraph engineering have already been mentioned. Another of the Board of Trade representatives was George Parker Bidder, famous in his day as a mathematical prodigy. So astonishing were his gifts, indeed, that it is worth reminding ourselves – in this age of electronic computers – that the human brain is still the most remarkable calculating machine in the known universe. Here are a few of Bidder's fully authenticated feats, in case anyone feels like trying to match them.*

At the age of nine, he was asked how long it would take to travel 123,256 miles at four miles a minute. He answered 21 days, 9 hours, 34 minutes – taking one minute to do the sum in his head. A year later he graduated to tougher problems; when asked how

* For an account of Bidder's performances, see W. W. Rouse Ball's *Mathematical Recreations and Essays*. The relevant section is reprinted in J. Newman's *World of Mathematics* (Vol. I).

many times a coach-wheel 5 feet 10 inches in circumference would revolve in running 800,000,000 miles it took him less than a minute to answer: '724,114,285,704 times with twenty inches left over'. The square root of 119,550,669,121 (345,761) took him only thirty seconds.

The interesting thing about Bidder was that, unlike many other lightning calculators, he was an able and intelligent man who became a very distinguished engineer, and was able to give an analysis of the methods he used. He also retained his powers throughout his life; when he was over seventy a friend happened to comment on the enormous number of light vibrations that must hit the eye every second, if there were 36,918 waves of red light in every inch and light travelled at 190,000 miles a second. 'You needn't work that out,' Bidder replied. 'The number is 444,443,651,200,000.'

When the committee presented its report in 1861, no less than 11,364 miles of submarine cable had been laid in various parts of the world – and only three thousand miles were operating. Most of the failures were due to bad design, workmanship or materials, the gutta-percha insulation being the chief cause of trouble. But the people who built the cables were not always to blame, as the farcical misadventures of the line from Sardinia to Algeria amply proved.

This cable was laid in the deepest water yet attempted, and the operation was in charge of John Brett, the pioneer of the English Channel cable. The French Government provided ships and navigation, and the net result was a superb example (not the last, one is tempted to add) of combined Anglo-French ineptitude.

On the first try, the cable was paid out so rapidly that the length provided was not sufficient for the job; Brett put the blame upon unexpected precipices on the Mediterranean sea-bed, whose presence had not been revealed by the French charts. His skipper remarked outright at the hearings, with a fine display of the Nelson spirit: 'It has been sounded by Frenchmen, and I have no reliance on their soundings.' But the Frenchmen were perfectly correct. The cable had not vanished into unknown abysses – it had merely run off the ship too quickly. Some years later another contractor hauled it up and, as he told the committee: 'We found huge masses of cable all tied together in a sort of Gordian knot, and *that* must have been one of Mr. Brett's precipices.'

On the second attempt, however, French seamanship was un-doubtedly to blame for the disaster. Listen to the indignant Mr. Brett's own account of the way in which success was missed by the small margin which in such a case meant total failure:

We passed all the great depths with perfect safety in the night, and arrived within ten miles of the land. At daylight in the early break of the morning, I saw the French vessel decor-ated with flags . . . according to the reckoning of the French captain, we are safe, and should land our cable with some miles to spare. They decorated their vessel as a triumph, and they were drinking champagne. . . . Our captain had given us warn-ing in the night that he thought we were drifting very much out of our course. This I communicated to Monsieur de la Marche, the officer appointed by the French Government, and he replied: 'We know what we are doing.' I thought very probably they did. By the next morning our captain said: 'Ask the French captain what his bearings are, to give us the latitude and longi-tude.' He did so, and it was found to be wrong, and to agree rather with the opinion of our captain. . . . I then begged our captain to give his figures, and I told him to get me a large board, and I chalked them up, I think about two feet long, so that there might be no mistake about it. I saw, on the part of the French officers, something like consternation; they retired to the cabin, and went through the calculation. When they returned they said: 'We find that we are wrong and you are right.'

And so Mr. Brett was left holding the end of his cable almost within sight of the African coast, while the chastened Monsieur de la Marche departed to Algiers in search of help.

I said: 'How long shall you be gone?' He said, 'Five or six days.' The question then, of course, became hopeless; we did the best we could to hold on for five or six days; we took the end of the cable, passed it round the vessel and stopped it so that no strain could come upon it. We sent, among many other messages by the cable, one to London to immediately put in hand from thirty to fifty miles of cable, and send it out to us as rapidly as possible. . . . We held on for five days and nights; the last two days there was a very violent strain, and a very heavy sea, the vessel pitching and rolling, but not yet breaking the cable. Most of the young clerks, who were Italians, were sick,* and I was alone on the deck when I saw a message coming in, saying that several miles of cable were in progress, and would be rapidly sent out to us. Within a few minutes afterwards the vessel gave one of the sudden plunges which had been repeated through the night, and the cable broke. . . .

After this débâcle, one might have thought that the third attempt would have been conducted with almost excessive care. Yet what happened this time was even more ridiculous; somebody forgot the difference between pounds and kilograms, and as a result the weight controlling the brake on the cable was only half what it should have been. So once again the bed of the Mediterranean was festooned with overlapping coils of expensive cable, and once again the supply ran out a dozen miles from the African coast. . . .

Yes, the time was undoubtedly ripe for a Royal Commission and for the replacement of the amateur engineers by professionals with a sound scientific training. Such men did exist – Professor Wheatstone, the Siemens brothers, Latimer Clark, for example – but there were far too few of them. The study of electricity was in such a primitive state that no agreed units existed; although it seems incredible, there was still no way of measuring resistance, voltage or current in quantities which would be understood by everybody. The very meanings of these terms were not generally understood; to most people, even those who worked with it, electricity was still a mysterious and almost occult power, and the way in which words such as 'intensity', 'tension', 'quantity', 'rate', 'velocity' were tossed about only added to the confusion. Voltage was defined in terms of so many Daniell's batteries; current by the deflections of whatever instrument the experimenter happened to be using. Volts, ohms and amps still belonged to the future, and one of the witnesses before the committee felt compelled to remark: 'It is a great pity that those who touch upon this question do not make themselves familiar with the laws of Ohm; it would set aside all those absurd discussions that take place.'

* The immunity of the British to seasickness is, of course, proverbial.

There can be little doubt that the man chiefly responsible for changing submarine cable practice from an esoteric art about as successful as rain-making to an exact mathematical science was William Thomson. Though he was still a relatively young man when he appeared before the committee, he was already famous and was listened to with great respect. One would like to have heard his Scots burr as he remarked, apropos of a proposal by one inventor to make a cable with the strengthening steel wires on the *inside*: 'It is about as well planned as an animal with its brains outside its skull.'* And on the unfortunate Dr. Whitehouse's patent relay, which was intended to write down the telegraphed messages automatically as they were received, he made this annihilating verdict: 'I find altogether two or three words and a few more letters that are legible, but the longest word which I find correctly given is the word "be".'

It would have been interesting to see how the Whitehouse relay coped with *'Dampfschiffahrtgesellschaft'* (steamship company), which, one witness complained, with a justifiable sense of grievance, was charged as a single word all over Europe according to telegraphic law.

Dr. Whitehouse was, of course, one of the chief witnesses. His appearance must have been somewhat embarrassing to all concerned, for half the board of inquiry consisted of his ex-colleagues whose fortunes and reputations he had been so instrumental in damaging.

It says much for Professor Thomson's nobility of spirit (if not for his powers of judgment) that he had attempted to shield Whitehouse from the wrath of the directors when the 1858 cable had failed, by stating that he was one of the company's most loyal and devoted servants. But what the directors thought of 'our late electrician' is made very clear by their reply to Thomson, written on August 25, 1858 – even before the cable had given up the ghost:

> Mr Whitehouse has been engaged some eighteen months in investigations which have cost some £12,000 to the company

---

* But here, as in the case of the earth's age, the Professor was completely wrong. The recipient of his scorn was just a century ahead of his time (see page 192).

and now, when we have laid our cable and the whole world is looking on with impatience . . . we are, after all, only saved from being a laughing-stock because the directors are fortunate enough to have an illustrious colleague . . . whose inventions produced in *his own* study – *at small* expense – and from *his own resources* are available to supersede the useless apparatus prepared at great labour and enormous cost. . . . Mr. Whitehouse has run counter to the wishes of the directors on a great many occasions – disobeyed time after time their positive instructions – thrown obstacles in the way of everyone, and acted in every way as if his own fame and self-importance were the only points of consequence to be considered. . . .

Yet despite all that had happened, Whitehouse refused to admit that he had made any mistakes, and threw up an imposing smoke-screen of experimental data to support his theories. He would not agree that his giant induction coils, with their thousands of volts, had been responsible for the breakdown of the cable, and put forward complicated arguments to 'prove' that signals produced by his coils could be transmitted more rapidly than those from batteries. He made one good point, however, when he tried to throw some of the blame on to Cyrus Field, who had refused to let him have the time he needed for his preliminary experiments. 'Mr. Field,' remarked the doctor, 'was the most active man in the enterprise, and he had so much steam that he could not wait so long as three months. He said, "Pooh, nonsense, why, the whole thing will be stopped; the scheme will be put back a twelve-month." '

It would have been better for the promoters if the scheme *had* been put back for twelve months. As it was, for lack of preparation it was delayed eight years.

One of the most remarkable, as well as most opinionated, characters to appear before the committee was Admiral Robert Fitzroy, FRS. The Admiral was a scientist of no mean repute, being a pioneer of meteorology and the founder of today's system of weather forecasting. However, what had given him a small but certain lien on immortality was the fact that, twenty years before, Captain Fitzroy had set out from England on one of the most momentous voyages of all time – the five-year cruise of HMS

Beagle, with a shy young scientist named Charles Darwin aboard. The results of that voyage had been published in three volumes, of which Fitzroy had written the first two and Darwin the last.

His career after that had been somewhat chequered. He had been an unsuccessful Governor of New Zealand, infuriating the colonists by supporting the rights of the natives. Tact was never his strong point, and at once time he even became embroiled in fisticuffs outside a London club. Perhaps his brilliant but un-balanced temperament was due to his somewhat unusual ancestry: he was a direct descendant of Charles II and the rapacious royal mistress Barbara Villiers, Duchess of Cleveland. (She could be tactless as well as rapacious; it once took all Charles's powers of persuasion to stop her combining *his* honeymoon with her con-finement.)

Only five years in the future, Admiral Fitzroy was to kill him-self in a fit of depression; but there was no sign of any uncertainty or lack of confidence as he threw off his ideas about insulators, the best routes for Atlantic cables, and methods of taking deep-sea temperatures. He wanted to coat the cable not with gutta-percha but with a flexible form of glass or 'vitreous substance' – a sugges-tion which had also been made by the Prince Consort. The Admiral pointed out that when it is immersed in water, glass is a much more pliable and manageable substance than when it is in air, and mentioned the extraordinary fact – which no one will believe until they try the experiment for themselves – that a sheet of glass held under water can be cut with an ordinary pair of scissors. Just how this would help submarine cables, Admiral Fitzroy did not explain.

He also ventured to *dissent entirely* (his italics) from the theory of *circuits formerly* held generally, and had the novel idea that there was no need for excessive care in making perfect joints when sections of cable were united. He thought that a simple snap-joint like that on a watch-chain would be sufficient, and that the ela-borate cleaning and soldering indulged in by electrical engineers was unnecessary.

Admirals, engineers, business men, cable contractors, scientists – for weeks and months they gave their views and experiences to the committee, while the Board of Trade clerks toiled to take down the hundreds of thousands of words of evidence. And then, of all the unexpected people to turn up in Whitehall, there arrived a colonel from Kentucky with a proposal that must have given the Atlantic Telegraph Company some anxious moments.

Colonel Tal P. Shaffner had built many of the first long-distance telegraphs in the United States, including, to use his own words, one 'from the Mississippi river to the western borders of civilisation'. ('That was through Kansas?' 'Yes, it was before Kansas was settled; it was at that time all occupied by Indians.')

The Colonel did not believe that a direct transatlantic cable was an economic proposition, and produced the complete transcript of all the messages passed through the 1858 line to prove his theory. In each direction, the cable had managed no more than a hundred words *per day* – and very few of those words had formed commercial messages, most of them being operating instructions or desperate attempts to find what was going wrong (see page 61). According to Shaffner's calculations, a two-thousand-mile-long submarine cable could never hope to pay for itself; even if it was in good electrical condition, it would be too slow to be of practical value.

Colonel Shaffner's alternative was to have a North Atlantic cable routed from Scotland to the Faroes, thence to Iceland, thence to Greenland, and finally to Labrador. The greatest length of submerged cable would then be no more than six hundred miles, and as all messages would be relayed on from the intermediate land stations as quickly as they were received, this six-hundred-mile section would determine the maximum working speed. It would be, as it were, the weakest link in the electrical chain; even so, it would be far superior to a continuous two-thousand-mile length, for theory indicated that it would be about ten times as fast.

The Colonel had spent a lot of time and money promoting his scheme, had carried out a survey of the route, and had obtained a concession from Denmark for the Faroes–Iceland–Greenland section of the proposed line. He had done this as early as 1854, and must therefore have been one of the very first men to realise the possibilities of transatlantic telegraphy. But though his scheme appeared attractive on paper, it involved building overland lines across some of the most desolate regions of the world, and laying submarine cables in waters infested with icebergs. Admiral Sir

James Ross, perhaps the greatest polar expert of the day, spoke out strongly against the dangers of drifting ice, and concluded that the direct southern route would be much safer and easier.

If the direct Ireland–Newfoundland route had not eventually succeeded, it seems possible that Colonel Shaffner's scheme might have been tried; indeed, today there is a cable from Scotland to Iceland *via* the Faroes, though it has never been continued to Greenland. It turned out that the Colonel underestimated the ability of the submarine engineers and was misled into thinking that all two-thousand-mile-long cables would perform as badly as the first. He was wrong; there was no need to use Greenland and Iceland as stepping-stones, and the gentleman from Kentucky lost his million-dollar gamble.*

Perhaps the most poignant evidence given before the committee was the account of Mr. Saward, secretary of the Atlantic Telegraph Company, of his efforts to obtain the £600,000 needed to lay a new cable. He reported sadly:

> I myself have personally waited upon nearly every capitalist and merchantile house of standing in Glasgow and in Liverpool, and some of the directors have gone round with me in London for the same purpose. We have no doubt induced a great many persons to subscribe, but they do so as they would to a charity, and in sums of corresponding amount. . . .

But the tide was slowly turning. Now that all the evidence had been aired and all the experts had given their opinions, the reasons for the earlier failures were apparent, and it was also clear how they could be avoided. The committee summarised its Herculean labours as follows:

> The failures of the existing submarine lines which we have described have been due to causes which might have been guarded against had adequate preliminary investigation been made into the question. And we are convinced that if regard be had to the principles we have enunciated in devising, manufacturing, laying and maintaining submarine cables, this class of

---

* An even bigger gamble on an alternative route was lost by Western Union (see page 97).

enterprise may prove as successful as it has hitherto been disastrous.

In other words: we've learned from our mistakes; now we can do the job. It was true enough; but success still lay five years in the future – and at the far side of yet another catastrophe.

CHAPTER 9

The Brink of Success

The first problem was to raise the money. It has been said that nothing is so nervous as a million dollars, and more than that had already disappeared into the Atlantic. Despite the technical evidence, and the fact that improved cables were now being laid in other parts of the world, Cyrus Field now had an exhausting battle before him. Between 1861 and 1864 he was continually travelling between England and America, trying to convince capitalists that next time there would be no failure. His success in his own country is summed up in this passage from his brother Henry's biography:

> The summer of that year [1862] Mr. Field spent in America, where he applied himself vigorously to raising capital for the new enterprise. He was honoured with the attendance of a large array of the solid men of Boston, who listened with an attention that was most flattering . . . there was no mistaking the interest they felt in the subject. They went still further; they passed a series of resolutions in which they applauded the projected telegraph across the ocean as one of the grandest enterprises ever undertaken by man, which they proudly commended to the confidence and support of the American public. *But not a man subscribed a dollar.*

In all fairness, it must be remembered that American big business now had some excuse for its lack of enterprise. The Civil War was raging and a country divided in twain had little energy or spirit for such a project. Moreover, relations between England and the North were still strained by the declaration of neutrality of May 14, 1861, in which the Confederacy had been granted the belligerent rights accorded to a sovereign nation. All Field's tact and remarkable powers of persuasion must have been exercised to the full as he shuttled back and forth across the Atlantic at a rate that would be notable even in these days of air transportation. By 1864, indeed, he had crossed the Atlantic no less than thirty-one times in the service of the company.

It took more than two years to get things moving again, and this time the project was largely financed and wholly carried out by Britain, only about a tenth of the capital coming from the United States. On April 7, 1864, the two contractors who between them had the greatest experience of submarine cable manufacture merged into a single company. Until that date cable core and insulation had been manufactured exclusively by the Gutta Percha Company, and the protective armouring had been largely made by Messrs. Glass, Elliott & Co. They now formed the Telegraph Construction and Maintenance Company, which still exists and which, with its associates, has now made more than 90 per cent of the submarine cables in the world.

The directors of the new firm, under their chairman John Pender, MP, were so confident of success that they immediately subscribed £315,000 of capital. Field himself had been responsible for raising no less than £285,000 from private investors, and with £600,000 in the bank the great project could now be started once again.

The next problem was to decide the design of the new cable. This time there was no headlong rush to get it manufactured and laid before proper tests had been carried out; everyone knew what that policy had cost. Scores of samples were examined and submitted to every conceivable electrical and mechanical ordeal; the design finally approved had a conducting core three times as large as the 1858 cable, and was much more heavily armoured. It could stand a breaking strain of eight tons, compared with only three for the previous cable, and was over an inch in diameter. Though it weighed one and three-quarter tons per mile, and was thus almost twice as heavy as its ill-fated predecessor, its weight when submerged in water was considerably less. This meant that the strain it would have to bear when being laid was also reduced, owing to the increased buoyancy. Indeed, ten miles of it could hang vertically in water before it would snap under its own weight; this was four times as great a length as could ever be suspended from a cable ship sailing across the North Atlantic, where there could never be more than two and a half miles of water beneath the keel.

Some thirty miles of more heavily armoured shore-end cable were produced by W. T. Henley of North Woolwich.

In every respect, the new cable was a vast improvement over any that had been built before. And yet, despite all the thought, skill and care that had gone into its construction, hidden within it were the seeds of future disaster.

By the end of May 1865, the 2,600 miles of cable had been completed. Its weight was seven thousand tons – twice that of the earlier cable which had required two ships to lay it. But this time, by one of history's fortunate accidents, the only ship in the world that could carry such a load was unemployed and looking for a job. In the Atlantic cable, the fabulous *Great Eastern* met her destiny and at last achieved the triumph which she had so long been denied.

This magnificent but unlucky ship had been launched seven years before, but had never been a commercial success. This was partly due to the stupidity of her owners, partly to the machinations of John Scott Russell, her brilliant but unscrupulous builder, and partly to sheer accidents of storm and sea.*

Seven hundred feet long, with a displacement of 32,000 tons, the *Great Eastern* was not exceeded in size until the *Lusitania* was launched in 1906 – forty-eight years later. She was the brain-child of Isambard Kingdom Brunel, the greatest engineering genius of the Victorian era – perhaps, indeed, the only man in the last five hundred years to come within hailing distance of Leonardo da Vinci. Brunel built magnificent stone and iron bridges which are standing to this day (the Clifton Suspension Bridge at Bristol is his most famous, though it was completed after his death) and threw superbly landscaped railways over most of southern England. He was as much an artist as an engineer, and the remorseless specialisation that has taken place since his day makes it impos-

---

* James Dugan's book *The Great Iron Ship* is a valuable and highly entertaining history of this wonderful vessel, but unfortunately repeats the legend that the skeleton of a riveter was found inside her double hull when she was broken up. This story is much too good to be true, and isn't. Dugan is also far too kind to Russell, whose evil genius not only laid a burden on the *Great Eastern* from which she never recovered, but undoubtedly contributed to the death of her designer. For this side of the story, see L. T. C. Rolt's important biography *Isambard Kingdom Brunel*.

sible that any one man will ever again match the range of his achievements.

Of these, the *Great Eastern* was his last and mightiest. Though she was *five times* the size of any other ship in the world, she was no mere example – as some have suggested – of engineering megalomania. Brunel was the first man to grasp the fact that the larger a ship, the more efficient she can be, because carrying capacity increases at a more rapid rate than the power needed to drive the hull through the water (the first depending on the cube of the linear dimensions, the second only on the square).

Having realised this, Brunel then had the courage to follow the mathematics to its logical conclusion, and had designed a ship that would be large enough to carry enough coal for the round trip to Australia. (Little more than a decade before, learned theoreticians had 'proved' that it was impossible for a steam-driven vessel even to cross the Atlantic.)

With her five funnels, six masts, and superb lines, the *Great Eastern* still remains one of the most beautiful ships ever built, though the absence of a superstructure makes her look a little strange to modern eyes. It is impossible to write of her without using superlatives; her fifty-eight-foot diameter paddle-wheels and twenty-four-foot screw have never been exceeded in size, and now never will be. This dual propulsion system made her the most manœuvrable ocean liner ever built; by throwing one wheel into reverse, she could rotate around her own axis as if standing on a turn-table.

By 1865, the *Great Eastern* had bankrupted a succession of owners and had lost well over £1,000,000. Put up to auction without reserve, the floating white elephant was knocked down for a mere £25,000 – about a thirtieth of her original cost. The buyers, headed by Daniel Gooch, chairman of the Great Western Railway, had already arranged with Cyrus Field to use the ship for laying the new cables; they were so confident that she could do it that they had offered her services free of charge in the event of failure.

To provide storage space for the huge coils of wire, three great tanks were carved into the heart of the ship. The drums, sheaves and dynamometers of the laying mechanism occupied a large part of the stern decking, and one funnel with its associated boilers had been removed to give additional storage space. When the ship sailed from the Medway on June 24, 1865, she carried seven thousand tons of cable, eight thousand tons of coal and provisions for five hundred men. Since this was before the days of refrigeration, she also became a sea-going farm. Her passenger list included one cow, a dozen oxen, twenty pigs, 120 sheep and a whole poultry-yard of fowl.

Many of the pioneers – one might say survivors – of the earlier expeditions were aboard. Among them were Field himself (the only American among five hundred Britishers), Professor Thomson, Samuel Canning, chief engineer of the Telegraph Construction and Maintenance Company, and C. W. de Sauty, the company's electrician. Commander of the ship was Captain James Anderson, but in all matters relating to the cable-laying Canning had supreme authority. Dr. Whitehouse was not aboard, even as a passenger.

The division of duties and responsibility on the voyage was somewhat unusual. The Atlantic Telegraph Company – represented primarily by Field, with Thomson as his expert adviser – was the customer for the job, but as the Telegraph Construction and Maintenance Company had put up more than half the capital, made the cable, and chartered the ship, it was not going to let its client interfere with the actual laying operations. So Field and Thomson were virtually passengers – though if the work did not come up the the specifications they had laid down they had, of course, the right to refuse it.

This time, with all the cable for the entire job in a single ship, there was no problem of splicing in mid-Atlantic; the *Great Eastern* would sail straight from Ireland to Newfoundland. Thanks to the presence on board of W. H. Russell (later to become Sir William), the famous war correspondent of the London *Times*, we have a complete record of the voyage, which was later published in a splendidly illustrated volume with lithographs by Robert Dudley.

The shore-end of the cable was landed at Foilhommerum Bay, a wild and desolate little cove five miles from Valentia Harbour. Hundreds of people had gathered to watch on the surrounding hills, which were dominated by the ruins of a Cromwellian fort. The scene was like a country fair, with pedlars and entertainers of

Top left: 9. Charles Bright, chief engineer of the Atlantic Telegraph
Company.

Top right: 10. Cyrus W. Field.

Bottom left: 11. Oliver Heaviside.

Bottom right: 12. William Thomson, Lord Kelvin.

Above: 13. Extruding the core of the Atlantic Telephone cable. (The elaborate equipment is necessary to maintain the polythene insulation at precisely the correct diameter.)

Below: 14. Winding the copper tapes on the cable.

all kinds making the most of the occasion. Nothing so exciting had ever happened before in this remote part of Ireland – but the crowds were disappointed in their hope of seeing the *Great Eastern*. There was no need, nor was it safe, for her to come inshore. She lay far out at sea while HMS *Caroline* brought the immensely heavy shore-end of the cable up to the coast and landed it over a bridge of boats.

The shore end was spliced aboard the *Great Eastern*, and on the evening of July 23, 1865, she turned her bows towards her distant goal. The escorting warships *Terrible* and *Sphinx*,

> which had ranged up alongside, and sent their crews up into the shrouds and up to the tops to give her a parting cheer, delivered their friendly broadsides with vigour, and received a similar greeting. Their colours were hauled down, and as the sun set a broad stream of golden light was thrown across the smooth billows towards the bows as if to indicate and illumine the path marked out by the hand of Heaven. The brake was eased, and as the *Great Eastern* moved ahead the machinery on the paying-out apparatus began to work, drums rolled, wheels whirled, and out spun the black line of the cable, and dipped in a graceful curve into the sea over the stern wheel.

As Russell remarked, 'happy is the cable-laying that has no history'. This laying was to have altogether too much. The next morning, eighty-four miles out, the testing instruments indicated an electrical fault at some distance from the ship. There was nothing to do but haul the cable aboard until the trouble was found.

At first sight, this would seem to be a fairly straightforward operation. But with the *Great Eastern* as she was now fitted out, it was anything of the sort. She could not move backwards and pick up the cable over the stern, where it was paying out, because she would not steer properly in reverse and there was also the danger of the cable fouling her screw. So the cable had to be secured by wire tackle, cut, and transferred the seven hundred feet to the bow. As Russell describes it:

> Then began an orderly tumult of men with stoppers and guy ropes along the bulwarks, and in the shrouds, and over the

F

boats, from stem to stern, as length after length of the wire rope flew out after the cable. The men were skilful at their work, but as they clamoured and clambered along the sides, and over the boats, and round the paddle-boxes, hauling at hawsers, and slipping bights, and holding on and letting go stoppers, the sense of risk and fear for the cable could not be got out of one's head.

It took ten hours to haul in as many miles of cable. When the fault was discovered, it was a very disturbing one. A piece of iron wire, two inches long, had been driven right through the cable, producing a short-circuit between the conducting core and the sea. It might have been an accident; but it looked very much like sabotage.

A new splice was made and paying-out started again. This time only half a mile had gone overboard before the cable went dead. Russell remarked despairingly:

Such a Penelope's web in twenty-four hours, all out of this single thread, was surely disheartening. The cable in the fore and main tanks answered to the tests most perfectly. But that cable which went seaward was sullen, and broke not its sulky silence. Even the gentle equanimity and confidence of Mr. Field were shaken in that supreme hour, and in his heart he may have sheltered the thought that the dream of his life was indeed but a chimaera. . . .

Luckily, the fault cleared itself; almost certainly it did not lie in the cable, but in the instruments or connections either at Valentia or aboard the ship. 'The index light suddenly reappeared on its path in the testing room, and the wearied watchers were gladdened by the lighting of the beacon of hope once more.'

On the fourth day, July 26, the *Great Eastern* ran into heavy seas which made it hard for the *Sphinx* and the *Terrible* to keep up with her. As she forged ahead at a steady six knots, hardly affected by the waves which battered her little escorts, the *Sphinx* slowly dropped astern and at last disappeared from view. This was a serious loss to the expedition, because owing to some oversight the *Sphinx* carried the only set of sounding gear.

The next two days were uneventful, and those aboard were

able to relax. The literary gentlemen produced a ship's newspaper containing local news and gossip, and it would be hard to improve on this standard of reporting:

Professor Thomson gave a lecture on 'Electric Continuity' before a select audience. The learned gentleman having arranged his apparatus, the chief object of which was a small brass pot, looking like a small lantern with a long wick sticking out at the top, spoke as follows: 'The lecture which I am about to give is on a subject which has ever been of great interest to the intellectual portions of mankind, and——' The luncheon bell ringing, the learned Professor was left speaking.

For all his erudition, Professor Thomson did not overawe his colleagues and they seem to have looked upon him with affection as well as respect. One of them remarked later: 'He was a thoroughly good comrade . . . he was also a good partner at whist when work wasn't on, though sometimes, when momentarily immersed in cogibundity of cogitation, by scientific abstraction, he would look up from his cards and ask "Wha played what?" '

The *Great Eastern* ploughed on across the waves, spinning out her iron-and-copper thread.

There was a wonderful sense of power in the Great Ship and in her work; it was gratifying to human pride to feel that man was mastering space, and triumphing over the wind and waves; that from his hands down into the eternal night of waters there was trailing a slender channel through which the obedient lightning would flash for ever instinct with the sympathies, passions and interests of two mighty nations.

On the afternoon of the seventh day, when eight hundred miles had been paid out, the alarms went again. The fault was close to the ship, so once more the cable was cut, secured by wire ropes, and hauled round to the bow for picking up.

Thousands of fathoms down we knew the end of the cable was dragging along the bottom, fiercely tugged at by the *Great Eastern* through its iron line. If line or cable parted, down sank the cable for ever. . . . At last our minds were set at rest; the iron rope was coming in over the bows through the picking-up machinery. In due, but in weary time, the end of the cable appeared above the surface, and was hauled on board and passed aft towards the drum. The stern is on these occasions deserted; the clack of wheels, before so active, ceases; and the forward part of the vessel is crowded with those engaged on the work, and with those who have only to look on . . . the two eccentric-looking engines working the pickup drums and wheels make as much noise as possible . . . and all is life and bustle forward, as with slow unequal straining the cable is dragged up from its watery bed.

It required nineteen hours of this nervous work before the fault was reached – though it would have taken only a few minutes if suitable equipment had been installed at the stern. The cable was respliced, paying-out commenced once more, and a committee of inquiry started to examine the faulty coils piled on deck.

Concern changed to anger when it was found that the cable had been damaged in precisely the same manner as before, by a piece of wire forced into it. 'No man who saw it could doubt that the wire had been driven in by a skilful hand,' Russell comments, and it was pointed out that the same gang of workmen had been on duty when the earlier fault occurred. The sabotage theory seemed virtually proved, and a team of inspectors was at once formed so that there would always be someone in the cable tank to kep an eye on the workmen.

On the morning of August 2, the *Great Eastern* had completed almost three-quarters of her task. Back at Valentia, the electricians were receiving perfect signals along the 1,300 miles of cable that had been paid out. They could even tell, by the gentle wavering of the galvanometer light-spot, exactly when the ship rolled, for minute currents were induced in the cable by the field of the 32,000-ton magnet as it swung back and forth on the face of the sea.

And then, without any warning, the signals stopped. The hours passed, and still no messages came along that thin thread leading out into the Atlantic. The hours grew into days; a week passed – then two. The *Great Eastern* and her escorts had vanished from human knowledge as completely as if the ocean had swallowed them up.

CHAPTER 10

Heart's Content

In England, the total disappearance of the telegraph fleet caused a storm of controversy and speculation. The *Great Eastern*, said the Jeremiahs, had probably broken her back on an Atlantic roller and gone down with all hands; she had been badly designed, anyway. (It was safe to say this, since Brunel had been in his grave for six years.) Arguments and counter-arguments raged in *The Times*, and in the midst of these the Atlantic Telegraph Company called an extraordinary general meeting. They announced not only their complete confidence in the cable that was now being laid, but stated they they would soon be asking for capital to make a *second* cable.

It was an act of extreme courage, and it was fully justified. Far out in the Atlantic the men of the *Great Eastern* were now proving that they would never accept defeat.

Cyrus Field had been one of the watchers on duty in the cable tank on the morning of August 2. About 6 a.m. there was a grating noise and one of the workmen yelled: 'There goes a piece of wire!' Field shouted a warning, but it did not reach the officer at the paying-out gear quickly enough. Before the ship could be stopped, the fault had gone overboard.

This time, it was not a complete short-circuit; the cable was usable, but no longer up to specification. Though Professor Thomson thought it could still transmit four words a minute – which would be enough to make it pay its way – Chief Engineer Canning decided not to take a risk. If he completed the cable, and the customer refused to accept it, his company would be ruined.

In any case, picking up a faulty section of cable was now a routine matter; the men had had plenty of practice on this trip. Canning had no reason to doubt that, after a few hours' delay, the *Great Eastern* could continue on the last seven hundred miles of her journey.

The cable was cut, taken round to the bows, and the hauling-up process began again. While this was going on, one of the workmen in the tank discovered some broken armouring wires on the piece of cable which had been lying immediately below the faulty section; the iron was brittle, and had snapped under the tremendous weight of the coils above it. This, said Russell, 'gave a new turn to men's thoughts at once. What we had taken for assassination might have been suicide!'

The *Great Eastern* was now over waters two thousand fathoms deep, though the exact depth was not known owing to the absence of the *Sphinx* with the only set of sounding gear. Unfortunately, she had neglected to signal Ireland that the cable was about to be cut; if this elementary precaution had been taken, much of the anxiety felt in England during the next two weeks would have been prevented.

From the start, the picking-up process failed to go smoothly. First the machinery gave trouble, then the wind made the *Great Eastern* veer round so that the cable did not come straight over the sheaves. It started to chafe against the ship, and when the picking-up machinery began to work once more, the strain on the cable proved too great for the weakened portion. 'The cable parted . . . and with one bound flashed into the sea. . . . There around us lay the placid Atlantic, smiling in the sun, and not a dimple to show where lay so many hopes buried.'

Now began a lonely mid-ocean battle which was to fire the imagination, and excite the admiration, of the world. Samuel Canning, despite the fact that the *Great Eastern* was not fitted with suitable gear, decided to fish for the cable two and a half miles down in the Atlantic ooze.

His men had grappled successfully for cables seven hundred fathoms down in the Mediterranean, but the depth here was three times as great. Even if the cable could be hooked, many doubted if it could stand the strain of being dragged up from such an abyss.

A five-pronged grapnel – 'the hook with which the giant Despair was going to fish for a take worth more than a million' – was attached to five miles of wire rope, and lowered over the side. It took more than two hours to sink to the sea-bed, but at last the slackening of the strain showed that it had hit bottom.

The *Great Eastern*, which had steamed several miles into the

wind, now shut off her engines and drifted under sail alone –
'the biggest sailing-ship the world will ever see', as Dugan points
out. All that night she moved silently down-wind, the grapnel far
below dragging through the deeper night of the ocean bed. Early
on the morning of August 3, the hooks caught in something, and
then began the nerve-racking business of hauling in the great
fishing-line with its unknown catch.

The grapnel-rope was most unsuitable for the work it was
doing (and which no one had ever imagined would be necessary)
because it was not in one continuous length but was made up of
two dozen sections each six hundred feet long, joined together by
shackles. These were the weak points of the rope, for when about
a mile of it had been drawn in over the bows a shackle parted and
rope, grapnel and the cable it had undoubtedly hooked fell back
to the sea-bed.

To add to Canning's difficulties, fog settled down and it was
impossible to take any observations which would fix the ship's
position. However, at noon on August 4 the sun providentially
broke through, a sight was taken, and it was found that the ship
was forty-six miles from the point where the cable had parted. A
large buoy was improvised and dropped overboard, anchored to
the sea-bed by three miles of the cable itself. Now the *Great
Eastern* had a fixed point from which she could work – a signpost
in mid-Atlantic, as it were.

No grappling was possible for the next two days, owing to an
unfavourable wind, but on August 7 Canning made his second try.
It was a repeat of the earlier performance; the cable was hooked
quickly enough, brought half-way to the surface – and another
shackle parted under the strain.

After this mishap, there was not enough wire left on board to
reach the bottom, so seven hundred fathoms of rope was spliced
on to make an improvised line. Bad weather and high seas held
up operations until the 10th; then the grapnel went overboard,
appeared to hook something, but came up too easily. When it
was hauled on deck, it was found that the line had twisted round
one of the flukes, effectively preventing the grapnel from doing its
job.

The fourth attempt was made the next day, and on the after-
moon of August 11 the cable was hooked again.

It was too much [wrote Russell] to stand by and witness the terrible struggle between the hawser, which was coming in fast, the relentless iron-clad capstan, and the fierce resolute power of the black sea. . . . But it was beyond peradventure that the Atlantic Cable had been hooked and struck, and was coming up from its oozy bed. What alterations of hopes and fears! Some remained below in the saloons, fastened their eyes on unread pages of books, or gave expression to their feelings in fitful notes upon piano or violin. . . . None liked to go forward, where every jar of the machinery made their hearts leap into their mouths. . . .

It was dark and raw that evening, and after dinner Russell left the saloon and walked up and down the deck under the shelter of the paddle-box.

I was going forward when the whistle blew, and I heard cries of 'Stop it!' in the bows, shouts of 'Look out!' and agitated exclamations. Then there was silence. I knew at once that all was over. The machines stood still on the bows, and for a moment every man was fixed, as if turned to stone. Our last bolt was sped. The hawser had snapped, and nigh two miles more of iron coils and wire were added to the entanglement of the great labyrinth made by the *Great Eastern* in the bed of the ocean.

There was a profound silence on board the Big Ship. She struggled against the helm for a moment as though she still yearned to pursue her course to the west, then bowed to the angry sea in admission of defeat, and moved slowly to meet the rising sun. The signal lanterns flashed from the *Terrible* 'Farewell!' The lights from our paddle-box pierced the night 'Good-bye! Thank you' in sad acknowledgement. Then each sped on her way in solitude and darkness.

The 1865 expedition had been yet another failure – but with a difference. It had proved so many important points that there could no longer be any reasonable doubt that a transatlantic cable could be laid. The *Great Eastern* had demonstrated, through her stability and handling qualities, that she was the perfect ship for the task; the cable itself was excellent, apart from the brittle armouring which could easily be improved – and, most important of all, it had been shown that a lost cable could be found and lifted in water more than two miles deep.

In a sense, therefore, the very failure engendered new confidence. Even so, one feels an admiration almost approaching awe for Cyrus Field and his partners as one considers their next step. They decided not merely to build and lay a completely new cable; when they had done *that*, they would go back and finish the one they had now three-quarters laid.

To circumvent some legal difficulties unsportingly raised by the Attorney-General on Christmas Eve, it was necessary to float a new company; it was also necessary to raise another £600,000. By early 1866, Field, assisted by Daniel Gooch and Richard Glass (managing director of the Telegraph Construction and Maintenance Company), had succeeded in doing this, and it is a tribute to the confidence that the public now had in them that when they needed £370,000 to make up the necessary capital of the new-born Anglo-American Telegraph Company it was subscribed in two weeks.

One thousand, nine hundred and ninety miles of fresh cable was at once ordered, incorporating various improvements over that used in the previous year. It was slightly stronger, yet lighter when submerged; more important, the brittle armouring had been replaced by a more ductile variety made from galvanised iron.

The arrangements aboard the *Great Eastern* had also been greatly improved. Machinery which could haul in the cable at the stern, so that it had no longer to be transferred to the bows, was the most important innovation. A method of continuous electrical testing had been devised, so that it was now impossible for several miles of cable to be paid out before a fault was detected. No less than twenty miles of wire rope, capable of standing a thirty-ton strain, was made to replace the inadequate grappling-line lost in mid-Atlantic. And the *Great Eastern* herself was given a much-needed spring cleaning, for 'her hull was encrusted with mussels and barnacles two feet thick, and long seaweed flaunted her sides'. The removal of these hundred of tons of marine growths must have added a couple of knots to her speed, and as the paddles had now been arranged to reverse independently, Samuel Canning had a vastly improved and more manœuvrable

cable-ship under his command when he set out from the Thames
on June 30, 1866.

This time, the Admiralty had been able to lend only one ship –
HMS *Terrible* – but the telegraph fleet was larger than last year
as the company had chartered the *Albany* and *Medway*. The latter
carried several hundred miles of last year's cable, as well as a
hundred miles of heavy shallow-water cable to lay across the St.
Lawrence.

As Sir W. H. Russell was now playing his more accustomed role
of war correspondent, we have to rely on other journalists for an
account of this trip. According to the *London News*, when the
massive shore-end of the cable – weighing eight tons to the mile
and looking like an iron bar – was landed at Valentia over a pon-
toon bridge of forty fishing boats,

> the ceremony presented a striking difference to that of last year.
> Earnest gravity and a deep-seated determination to repress all
> show of the enthusiasm of which everyone was full, was very
> manifest. . . . There was something far more touching in the
> quiet and reverent solemnity of the spectators than in the
> slightly boisterous joviality of the peasantry last year. . . . The
> old crones in tattered garments who cowered together, dudheen
> [clay-pipe] in mouth, their gaudy shawls tightly drawn over
> head and under the chin – the barefoot boys and girls – the
> patches of bright colour furnished by the red petticoats and
> cloaks – the ragged garments, only kept from falling to pieces
> by bits of string and tape. . . .

Such were the inhabitants of this poverty-stricken region; one
can understand their interest in the magic wire which led to the
land where so many of their countrymen had gone in search of a
better life.

Thirty miles out from land, the *Great Eastern* was waiting for
the special section of heavy shore cable, designed to resist both
angry seas and dragging anchors, to be brought out to her. When
the end was lifted aboard,

> quick, nimble hands tore off the covering from a foot of the
> shore-end and of the main cable, till they came to the core;
> then, swiftly unwinding the copper wires, they laid them

together, twining them as closely and carefully as a silken braid. Then this delicate child of the sea was wrapped in swaddling-clothes, covered up with many coatings of gutta-percha, and hempen rope, and strong iron wire, the whole bound round and round with heavy bands, and the splicing was complete.

And so on Friday, July 13, 1866, the *Great Eastern* sailed again from Valentia Bay. Those who disliked the date were reminded that Columbus sailed for the New World on a Friday – and also arrived on one.

At a steady and uneventful five knots, Brunel's masterpiece plodded across the Atlantic, paying out the cable with clockwork regularity. The only incident on the entire fourteen days of the voyage was when the cable running out from the tank caught in an adjacent coil, and there was a tangle which caused a few anxious moments before it was straightened out.

In England, where the progress of the expedition was known at every minute, excitement and confidence mounted day by day. In the United States, however, it was different, for there was no news of what was happening, nor could there be until the ships actually arrived – if they did. Some spectators were waiting hopefully in Newfoundland, but, as Henry Field remarks:

not so many as the last year, for the memory of their disappointment was too fresh, and they feared the same result again.

But still a faithful few were there who kept their daily watch . . . it is Friday morning, the 27th July. They are up early and looking eastwards to see the day break, when a ship is seen in the offing. Spy-glasses are turned towards her. She comes nearer – and look, there is another and another. And now the hull of the *Great Eastern* looms all glorious in that morning sky. They are coming! Instantly all is wild excitement. The *Albany* is the first to round the point and enter the bay. The *Terrible* is close behind. The *Medway* stops an hour or two to join on the heavy shore-end, while the *Great Eastern*, gliding calmly in as if she had done nothing remarkable, drops her anchor in front of the telegraph house, having trailed behind her a chain of two thousand miles, to bind the old world to the new.

*   *   *

No name could be more appropriate than that of the landing-place – Heart's Content. It was

a sheltered nook where ships may ride at anchor safe from the storms of the ocean. It is but an inlet from the great arm of the sea known as Trinity Bay, which is sixty or seventy miles long, and twenty miles broad. (The old landing of 1858 was at the Bay of Bull's Arm, at the head of Trinity Bay, twenty miles above.)

Heart's Content was chosen now because its waters are still and deep, so that a cable skirting the north side of the Banks of Newfoundland can be brought in deep water almost till it touches the shore. All around the land rises to pine-crested heights; and here the telegraph fleet, after its memorable journey, lay in quiet, under the shadow of the encircling hills.

As soon as the *Great Eastern* was anchored, Captain Anderson and his officers went ashore and attended a service in the local church, where the preacher, not very tactfully, addressed them on the text 'There shall be no more sea'. – An edict, according to Kipling's *The Last Chantey*, which caused such a protest in Heaven that it had to be rescinded so that

> . . . the ships shall go abroad
> To the Glory of the Lord
Who heard the silly sailor-folk and gave them back their sea!

The triumph was marred by a slight annoyance; the St. Lawrence cable had been broken, and so there was a delay of two days before the telegraph connection could be completed to the United States. It was not until the morning of Sunday, July 27, that New York received the message:

'Heart's Content, July 27. We arrived here at 9 o'clock this morning. All well. Thank God, the cable is laid and in perfect working order. Cyrus W. Field.'

On the first day of operation, the new cable earned £1,000. At last the sea was giving back the fortunes it had swallowed. Field was thankful, but he had not yet attained his own Heart's Content. There was still one more job to be done; neither he nor the men of the *Great Eastern* could rest while out there in the Atlantic, seven hundred miles away, the broken end of last summer's cable lay in the icy darkness beneath its tangle of grappling-irons.

CHAPTER 11

Battle on the Sea-bed

So eager was the telegraph fleet to resume its battle with the Atlantic that the *Albany* and *Terrible* left Newfoundland almost immediately, and were back in mid-ocean within five days. It took a little longer to get the *Great Eastern* ready for sea again; six hundred miles of the 1865 cable and several thousand tons of coal had to be loaded aboard. The cable had come over in the *Medway* – the coal in a small fleet of colliers sent from England, one of which had foundered on the way.

On August 9 the *Great Eastern* and the *Medway* put to sea again, making a rendezvous with the *Albany* and *Terrible* three days later. The two ships had already marked the line of the lost cable with buoys and then, 'like giant sea-birds with folded wings, sat watching their prey'. The *Albany* had made a brave attempt to lift the cable with her own gear, and had indeed succeeded in hooking it and bringing it some way to the surface before it slipped back into the depths again.

The *Great Eastern* stood off a few miles from the line of buoys and lowered her grapnel on its two-inch-thick wire rope. When it had reached the bottom, she started to drift downwind towards the buoys, in the expectation that, after one or two sweeps, the cable would be hooked.

But is was not as easy at that. Again and again the *Great Eastern* drifted back and forth without result. Sometimes Cyrus Field would go to the bow, sit on the rope, and tell by its vibrations that the grapnel was dragging on the sea-bed two miles below. The ocean floor

proved to be generally ooze, a soft slime. When the rope went down, one or two hundred fathons at the end would trail on the sea floor; and when it came up, this was found coated with mud, very fine and soft like putty, and full of minute shells. But it was not all ooze on the bottom of the sea, even on this telegraphic plateau. There were hidden rocks lying on that broad plain. Sometimes the strain on the dynamometer would suddenly go up three or four tons, and then back again, as if the grapnel had been caught and broken away. Once it came up with two of its hooks bent, as if it had come in contact with a huge rock. . . .

It was not until late on the night of August 16 that the cable was hooked. By morning it had been brought to the surface, and a great cheer went up as the company's lost treasure came back into the light of day.

We were all struck with the fact that one half of it was covered with ooze, staining it a muddy white, while the other half was in just the state in which it left the tank, which shows that it lay in the sand only half embedded. The strain on the cable had given it a twist, and it looked as if it had been painted spirally black and white. . . .

Unfortunately, the cable had been weakened by the strain of lifting, and before it could be properly secured it snapped and fell back into the sea. It had been visible, as if to tantalise its seekers, for just five minutes.

It almost seemed as if the expedition's luck had turned against it once more. Day after day, grappling continued without success. Sometimes the cable would be hooked, but each time it broke. There was one bitter disappointment when the *Albany* did succeed in bringing an end aboard – only to find that it came up much too easily. She had merely caught a fragment two miles long, broken off by the earlier attempts at lifting.

Supplies were now beginning to run out, and the *Terrible*, which had been at sea for a month, had to return to base. Her crew were already on half rations, but her captain sailed away with reluctance, 'mourning his fate, like a brave officer who is ordered away in the midst of a battle'.

By the end of August, the remaining ships decided to try new tactics. They moved a hundred miles eastwards, into slightly shallower water, and the grapnel went over for the *thirtieth* time. Again the cable was caught – but this time it was lifted only half-way to the surface and buoyed there, while the *Great Eastern* moved off a few miles and grappled again. Now the cable was secured at two points, the strain on it was not so great, and after twenty-four hours' patient hauling the great prize was at last brought abroad.

At once the end was taken down to the electricians' room, the core was stripped, and the instruments were connected to see if Ireland was still at the end of the line. It was possible that all this weary angling had been in vain if the cable had developed a fault somewhere along its length, or had been in any way injured during its year of submergence.

A silent crowd gathered to await the verdict; of the many tense moments the *Great Eastern* had known, this was perhaps the most dramatic. As Henry Field describes it:

> . . . the accustomed stillness of the test-room is deepened; the ticking of the chronometer becomes monotonous. Nearly a quarter of an hour passed, and still no sign! Suddenly the electrician's hat is off, and the British hurrah bursts from his lips, echoed by all on board with a volley of cheers. . . . Along the deck outside, throughout the ship, the pent-up enthusiasm overflowed; and even before the test-room was cleared, the roaring bravoes of our guns drowned the hurrahs of the crew, and the whizz of the rockets was heard rushing high into the clear morning sky to greet our consort-ships with the glad intelligence. . . .

The scene at the other end of the cable was less exuberant yet, in its way, quite as moving. It has been beautifully described in the *Spectator*:

> Night and day, for a whole year, an electrician has always been on duty, watching the tiny ray of light through which signals are given, and twice every day the whole length of wire – one thousand two hundred and fifty miles – has been tested for conductivity and insulation. . . . The object of observing the ray of light was not of course any expectation of a message, but simply to keep an accurate record of the condition of the wire. Sometimes, indeed, wild incoherent messages from the deep did come, but these were merely the results of magnetic storms and earth-currents, which deflected the galvanometer rapidly, and *spelt the most extraordinary words and sometimes even sentences of*

nonsense. Suddenly, one Sunday morning, while the light was being watched by Mr. May, he observed a peculiar indication about it, which showed at once to his experienced eye that a message was at hand. In a few minutes the unsteady flickering was changed to coherency . . . and the cable began to speak the appointed signals, instead of the hurried signs, broken speech and inarticulate cries of the illiterate Atlantic. . . . After a long interval in which it had brought us nothing but the moody and often delirious mutterings of the sea, the words 'Canning to Glass' must have seemed like the first rational words uttered by a high-fevered patient, when the ravings have ceased and his consciousness returns.

The splice was made, and the *Great Eastern* turned once more towards the west. This time, the whole world could follow her progress. She could talk to Europe through the cable she was reeling out; Europe could speak to America through the one she had already laid. There were scenes of tremendous excitement at Heart's Content when, despite an encounter with a severe storm, the *Great Eastern* brought in her second transatlantic cable only four weeks after she had arrived with her first.

The long and weary battle was ended. From that day to this, America and Europe have never been out of touch for more than a few hours at a time.

The queen of the seas, her triumph complete, turned eastwards again. The first ship that would exceed her in size was still forty years in the future, and many memories must have passed through Cyrus Field's mind as he said good-bye to his friends. 'As he went overboard, "Give him three cheers!" cried the commander. "And now three more for his family!" The ringing hurrahs of the gallant crew were the last sounds he heard as . . . the wheels of the *Great Eastern* began to move, and that noble ship, with her noble company, bore away for England.'

The word 'noble' was appropriate. Queen Victoria, who together with the Prince Consort had always been extremely interested in the project, immediately conferred knighthoods upon Thomson, Glass and Canning, as well as upon Captain Anderson. Daniel Gooch became a baronet, as did Lampson, deputy chairman of the Atlantic Telegraph Company. Perhaps one should not

Above: 15. The last stage in cable manufacture: winding the jute wrapping.

Below: 16. Loading cable onto HMTS *Monarch* at the Thames-side factory of Submarine Cables Ltd.

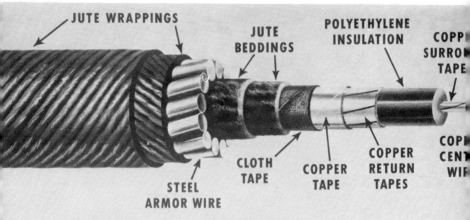

JUTE WRAPPINGS
JUTE BEDDINGS
POLYETHYLENE INSULATION
COPP[
SURRO[
TAPE[

STEEL ARMOR WIRE
CLOTH TAPE
COPPER TAPE
COPPER RETURN TAPES
COP[
CEN[
WIF[

Submarine Telephone Cable

Above: 17. Cutaway section of the submarine cable.

Below: 18. The four types of cable used on the Atlantic telephone. They differ only in the armouring: deep-sea section on right, shore end on left.

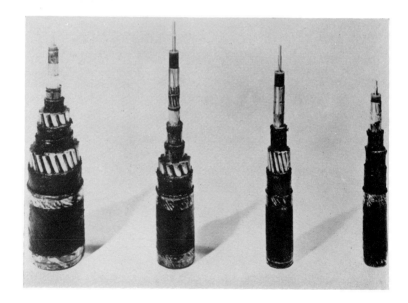

be carping, as there was honour enough for all, and all deserved it; but it seems a little odd that a higher distinction was given to the men who provided the money than those who did the work.

Just how well they had done it was proved by a famous experiment carried out at Valentia by the electrical expert Latimer Clark a few weeks after the second cable was laid. He gave orders that the Newfoundland ends of the two cables should be connected together, thus providing a submarine circuit more than four thousand miles in length. And through this four thousand miles he was able to send clear signals, using as a source of power a battery made out of a lady's silver thimble containing a few drops of acid. There is, unfortunately, no record of what Dr. Whitehouse, whose five-foot induction coils were now gathering dust, thought of this final refutation of his brute-force theories.

It is a good wind that blows nobody some ill; the achievement of transatlantic telegraphy had been a death-blow to a mighty and now forgotten enterprise on the other side of the globe. Colonel Shaffner's project for a line via Greenland has already been mentioned in Chapter 8; this came to nothing, but an alternative and much more grandiose scheme was well under way when the *Great Eastern* returned to England in triumph.

This was the so-called 'Overland Line to Europe', a telegraph circuit which would run from America through British Columbia, Alaska, Siberia and Russia. Instead of two thousand miles of submarine cable there would be sixteen thousand miles of land-line; the narrow Bering Straits, of course, presented no serious problem at all.

Convinced that the Atlantic cable, even if it could be laid, would be uneconomic, the Western Union Company started work on the Overland Line in March 1864. It chartered ships, organised land expeditions and carried out surveys of the barren and uninhabited regions through which the telegraph would run. Three years of toil and $3,000,000 of cash were poured into the project. The engineers and workmen were still full of enthusiasm, despite the hardships they had endured, when a passing ship brought the news to their remote Siberian camp that there was now not one, but two, cables linking Europe with America.

Robert Luther Thompson, in his historical survey *Wiring a Continent*, has an entertaining account of the way in which the great project collapsed:

They opened a sort of international bazaar and proceeded to dispose of their surplus goods upon the best terms possible. They cut the price of telegraph wire until that luxury was within the reach of the poorest Korak family. They glutted the market with pick-axes and shovels . . . which they assured the natives would be useful in burying the dead, and then threw in frozen cucumber pickles which they warranted to fortify the health of the living. . . . They taught the natives how to make cooling drinks and hot biscuits, in order to create a demand for their redundant lime-juice and baking powder. They directed all their energies to the creation of artificial wants in that previously happy and contented community. . . . But the market at last refused to absorb any more brackets and pick-axes; telegraph wire did not make as good fish-nets and dog-harnesses as the Americans confidently predicted; and lime-juice, even when drunk out of pressed crystal insulators, beautifully tinted with green, did not seem to commend itself to the aboriginal mind. . . .

So, like a defeated army leaving its stores scattered over the battlefield, the workmen and engineers trickled home. Yet though they had failed in one enterprise, they had achieved something equally important. They had opened up British Columbia, and they had drawn the attention of the United States to the hitherto ignored region known as 'Russian America'. In the year that the Overland Line was abandoned, the territory through which it would have passed was purchased from Russia, largely at the instigation of Secretary of State Seward. He had to face almost as much opposition from Congress as when he had backed Cyrus Field and the Atlantic cable ten years before; but it is now generally agreed that the United States had a good bargain when it bought Alaska for $7,200,000.

As for Cyrus Field, his life's great work was now completed, though he was still only forty-seven. How he saw that work can best be shown by his own words, from a speech he gave at a banquet arranged in his honour by the New York Chamber of

Commerce on November 15, 1866. They are still as timely as when they were delivered more than a century ago:

As the Atlantic Telegraph brings us into closer relations with England, it may produce a better understanding between the two countries. Let who will speak against England – words of censure must come from other lips than mine. I have received too much kindness from Englishmen to join in this language. I have eaten of their bread and drank of their cup, and I have received from them, in the darkest hours of this enterprise, words of cheer which I shall never forget; and if any words of mine can tend to peace and goodwill, they shall not be wanting. . . . America with all her greatness has come out of the loins of England, and though there have sometimes been family quarrels, still in our hearts there is a yearning for our old home, the land of our fathers; and he is an enemy of his country and of the human race, who would stir up strife between two nations that are one in race, in language and in religion.

The year after the cable was opened, Congress made amends for its earlier treatment by giving Field a unanimous vote of thanks and awarding him a gold medal – though owing to the stupidity of a Goverment clerk it took several years to reach him. He had achieved fame in his own country, which he also served in many other ways. After the Civil War, he had helped to smooth the controversies that developed between England and the United States, and when President Garfield was assassinated in 1881 Field started the fund which raised $362,000 for his dependants. A few years later he tried to launch a subscription to aid Grant, then ill and in financial difficulties, but a letter from the proud old general deterred him.

These acts were proof of Field's generosity, and it is sad to relate that in his own later years monetary and personal troubles came upon him. When he was over seventy, he discovered that some of his business colleagues had depreciated his stocks, and that only a few thousands were left of his once considerable fortune. Yet two years before his death in 1892, at the age of seventy-three, he had the happiness of celebrating his golden wedding surrounded by seven of his children and many more grandchildren.

There are few men who have achieved so much in the face of such overwhelming obstacles. And he had done this without the ruthless brutality which characterised many of the other great financiers of the epoch. According to one of his contemporaries, he was both visionary and chivalrous, but because of his own frankness he sometimes underestimated the selfishness and ingratitude of others.

Sir William Thomson, on the other hand, was to go on to yet greater fame and achievement once the 1866 cable was safely laid. Four years later he succeeded in making an instrument which would automatically record even the feeblest of signals, so that there would be a permanent record on tape. It is hard to imagine what the strain on the telegraph clerks must have been before this problem was solved; they had to sit for hours on end, watching a spot of light jiggling to and fro – and if they took their eyes off it for a moment a letter or a word would be lost. It is hardly surprising that one of the witnesses at the 1861 inquiry stated that quarrelling between the operators was a serious cause of lost time, the clerks sometimes becoming so irritable that they would refuse to work.

In his earlier mirror galvanometer, Thomson had used a beam of light to provide a frictionless, weightless pointer; now he produced a frictionless pen. The heart of his siphon recorder was a very fine glass tube, bent into a U. One end dipped into an ink-bottle, while the other was held a fraction of an inch above a moving paper tape. There was no physical contact – and hence friction – between pen and paper; the ink was electrified so that the paper attracted it and it jumped the gap without producing any drag as the pen wrote its dots and dashes in the form of a continuous wavering line.

This durable instrument was used until well into this century; indeed, its direct descendants may still be seen standing in cable offices even today.

Thomson was now doing well out of his inventions; it was no coincidence that in the year that he produced the siphon recorder he also bought his yacht *Lalla Rookh*. After that he spent much of his spare time at sea – frequently testing more inventions. One of the most important of these was a new method of taking deep-sea soundings, using piano wire instead of the hemp ropes that had previously been employed. Soundings could now be taken while a ship was under way, for the thin wire was much less affected by the drag of the water. Thomson was no theoretical seaman; he once navigated across the Bay of Biscay using only his new sounding machine, which even the modern echo-sounder has not yet made obsolete. And he revolutionised the design of the mariner's compass, notwithstanding the usual dogged opposition of HM Lords of the Admiralty.

It is also pleasant to record that Thomson's telegraphic activities brought him personal happiness as well as riches; he was a widower when he met his second wife while on an expedition to lay a cable to South America. Throughout the remainder of the nineteenth century his fame steadily increased. In 1892 he became Lord Kelvin of Largs, and when he died in 1907 his long life had spanned the almost unimaginable gulf between the first steam locomotive and the first aeroplane. Yet all that time – though he never knew it – he had been engaged on a hopeless quest. To the very last he had striven to understand the universe in mechanical terms; it was almost as if he hoped that one day science might be able to produce an engineering blue-print of the atom. Today we know the futility of that dream; only a dozen years after Kelvin of Largs was laid in Westminster Abbey, the first successful test of the Theory of Relativity showed that the universe was a far stranger place than he had ever imagined.

CHAPTER 12

Girdle Round the Earth

With two submarine cables in operation across the Atlantic, there could be no further doubts about the future of this method of communication. From 1866 onwards, cables spread swiftly across the seas and oceans of the world. In 1869 the *Great Eastern* laid her third Atlantic cable – this time one almost three thousand miles long, between France and the United States.

In 1870, Britain and India were linked – again by the *Great Eastern* – and the failures of a decade before were redeemed. Previously, it had often taken a week for telegrams to reach India by the overland route, and they had frequently arrived in an indecipherable condition after having been relayed by clerks of many different nationalities. The submarine cable avoided all language and political problems, and London could now get a reply from Bombay within a matter of minutes.

A year later, in 1871, Australia was reached by way of Singapore, and in 1874 the first cable was laid across the South Atlantic, connecting Brazil to Europe *via* the islands of Madeira and St. Vincent. Before she had finished her career, the *Great Eastern* was to lay five Atlantic cables, the spanning of the Pacific, however had to await the coming of the twentieth century.

Few things are duller than a record of steady and uninterrupted progress; when the telegraph fleet anchored in Heart's Content on July 27, 1866, the adventurous, pioneering days were over, and with them the excitement they engendered. From then onwards the story was largely one of straightforward commercial development, as is proved by the fact that when the century ended no less than fifteen cables had been laid across the North Atlantic. The 1865 and 1866 cables lasted for five years before they had to be repaired, but sections of the 1873 cable from Ireland to Newfoundland are still in service today after more than a century of operation. Once a well-designed cable has been safely laid in deep water, there are very few things that can stop it working, and apart from accidents its natural span of life is measured in decades. But as we shall see in the next chapter, accidents can and do happen even on the bed of the ocean.

The very longevity of a submarine cable – almost unprecedented for any technical device – has to some extent acted as a brake on the progress of long-distance telegraphy. It costs a million pounds or more to lay a trans-ocean cable, and even if somebody invents a much better one ten years later, there is little incentive to abandon a piece of equipment which may give half a century of reliable operation.

As we shall see later, in a hundred years there have been only three major improvements in the design of submarine cables; it is the transmitting and receiving equipment which has changed out of recognition, from the days when a telegraph clerk tapped a key at one end of the line, and another watched a needle or a spot of light dancing back and forth at the other.

In the days of manual operation, messages had to be taken down and retransmitted at every station along the cable; there might be six or more repetitions on a long line, with all the possibility of error and delay that they introduced. There was obviously an urgent need for a device which would automatically receive signals as they came from one section of a cable and pass them on again as good as new into the next length, however weak and distorted they might have been when they arrived. Merely amplifying or magnifying the incoming signal (even if anyone had known how to do this back in the 1870s) was not enough, because that would also amplify the imperfections the signal had picked up during its journey along the line. After two or three stages of this, it would be impossible to tell a 'dot' from a 'dash'.

What was wanted was an instrument which would carry out the same functions as the human relays – the telegraph clerks. They recognised whether the incoming signal was a dot or a dash, and sent out a fresh signal which was as good as new – if the identification had been properly made.

It was not until the 1920s that a reliable instrument capable of doing this made its appearance. Its name – the 'regenerator' – describes its function perfectly. It scrutinises each impulse that comes along the line, decides whether it is a dot or a dash and then makes a fresh dot or dash which is not a mere copy of the incoming one but a brand-new article.

At this point, it should be explained that though the telegraph companies still use Morse code (as well as other codes) the terms 'dot' and 'dash' are now somewhat misleading for the two basic elements of the code are no longer distinguished by their length. When a Boy Scout sends Morse by sounding a buzzer or flashing an electric torch, he is dealing in the old-fashioned dots and dashes, and an SOS comes out as:

Dit Dit Dit   Dā Dā Dā   Dit Dit Dit

or

Flick Flick Flick   Flāsh Flāsh Flāsh   Flick Flick Flick

the dashes lasting about twice as long as the dots.

This is uneconomical and inconvenient for automatic working, and in the normal cable code a dot lasts exactly as long as a dash. The two are distinguished by making a dot a pulse of negative current, a dash a pulse of positive current. So an SOS in cable code would be

$$- - -\quad + + +\quad - - -$$

or in terms of a moving needle or spot of light

Left Left Left   Right Right Right   Left Left Left

When a message is sent, for example, from London to Hong Kong, it may be passed through a dozen regenerators in series before it reaches its destination approximately one second later. At each intermediate station – Porthcurno (Cornwall), Carcavelos (Spain), Gibraltar, Malta, Alexandria, Suez, Port Sudan, Aden, Seychelles, Colombo, Penang, Singapore – all the dots and dashes will have been scrutinised and repeated almost faster than the eye can watch the machines operating. And today there is not even a human operator at the end of the line; the message is automatically printed on paper tape that can be glued to a cablegram form and delivered to the addressee.

One of the most important developments in the early days of submarine telegraphy was a means of sending signals simultaneously in each direction – a useful trick that almost doubled the working capacity of a circuit. This is known as duplex operation, and like many other electrical feats seems somewhat miraculous until you are shown how it is done. The secret lies in making the receiver at your end of the line insensitive to the impulses you are transmitting, while still capable of picking up the incoming signals. When you stop to think about it, you will realise that Nature has done exactly the same thing with your sense of hearing. When you speak, you cannot hear your own voice except as a very faint ghost of the original – as is amply proved by the fact that nobody ever recognises a recording of himself. But you can hear anyone else speaking, even when you are talking at the same time. So can telegraph instruments working on a duplex circuit.

The next development – which in some ways seems even more remarkable – was *multiplex* working, in which several messages could be sent in the same direction at the same time. Thus one might have a single cable carrying eight separate messages simultaneously – four in each direction. Once again, the trick was simple, though as is usually the case the practical application was much harder than the theory. A rapidly operating switch connected the line to each of the sending instruments in turn; each would have the use of the line for a fraction of a second, and then hand over to the next sender. At the far end, the receiving instruments would be switched in and out at exactly the same rate. It is precisely as if four people with only one telephone between them took turns to speak to four people at the other end of the line, talking for say one minute in strict rotation.

By the use of such techniques, it was possible by the 1930s to send up to four hundred words per minute across the most up-to-date of the Atlantic cables. This is just about a hundred times better than the 1858 cable could do, even at those rare moments when it was working properly.

One of the results of the rapid extension of the submarine telegraph system over the face of the globe was that some very odd and out-of-the-way places, which had previously been no more than names on the map – and sometimes not even that – suddenly became of great commercial and strategic value. Since the rate of working of a cable falls rather rapidly as its length increases, it is important to keep the sections as short as possible, linking them together with the relays or regenerators already described.

By bad luck, there is no suitable island in mid-Atlantic for the convenience of the cable (and airline) companies. The first direct circuit between England and the United States (TAT 3) was not established until 1963. For reasons which are better apparent by looking at a globe rather than a map, for a long time the cables preferred to take the short northern route *via* Newfoundland, or else go south and make a stop at Fayal in the Azores before continuing their journey.

Such remote pin-points of land as Ascension in the South Atlantic, Fanning, Guam and Midway in the Pacific, and Cocos in the Indian Ocean, have become cross-roads of world communications simply by virtue of their geographical positions. More than once this has turned a coral atoll that would seem to be of no use to anyone into a major military objective.

The classic case in the first world war was that of the Cocos Islands, junction of the cables from South Africa, the East Indies and Australia. On November 9, 1914, the German cruiser *Emden* put a landing party ashore on the Cocos to destroy the station and cut the cables. It was a Pyrrhic victory; before the line went dead the station transmitted the alarm which brought the Australian cruiser *Sydney* racing to the attack. The *Emden* was sunk, in the first major naval engagement of the 1914–18 war – which gave as big a boost to the morale of the Allies as did the Battle of the River Plate twenty-five years later

In the second world war, the Japanese tried a repeat performance when one of their battleships shelled the island on March 3, 1942. Perhaps remembering the fate of the *Emden,* the raider did not stay to see if the cable station had been destroyed, and thus the British were able to work a bluff which they maintained until the end of the war. The circuit was still intact, but radio messages were sent to the adjacent stations, in plain and interceptible English, ordering them to destroy their instruments because Cocos had been put permanently out of action. At the same time, of course, orders were sent by cable telling them to do nothing of the sort but to ignore the radio instructions. And so the Japanese never bothered defenceless Cocos Island again until the Far Eastern war was over.

Besides the cable stations, the cables themselves have been prime objectives during both world wars. In 1939, the Germans possessed only two cables of their own – one from Emden to the Azores, the other from Emden to Lisbon. Both were cut with neatness and dispatch before the war was twenty-four hours old, in the first move of a battle of wits which was to continue on the sea-bed for the next six years.

It is not hard to cut a cable if you know its approximate position; all you have to do is to trawl at right-angles until you hook it. You then have a choice of cutting it where it lies, using the special grapnels that have been designed to do this, or hauling it up to the surface and perhaps reeling it in for your own use.

This is obviously only possible if you have command of the seas and can protect your cable-ship from enemy attack. If this is not the case, you will have to do the job from a submarine; such an operation was actually carried out in 1945 when the XE4, a British midget submarine, cut the Saigon–Singapore and Saigon–Hong Kong cables.

This feat – one of the most remarkable operations of the war – was carried out by two divers, Sub-Lieutenants K. M. Briggs and A. K. Bergius. Their tiny four-man craft was towed to the area by a larger submarine, and then proceeded on its own course a few feet above the sea-bed, dragging a grapnel. After several runs the cables were hooked, and the divers left the submarine carrying power-operated cutters. It was a risky operation, for they were working in a strong current at a depth where oxygen poisoning is liable to occur. But they cut the cables successfully, and brought back sections as souvenirs.

To cause the maximum annoyance to the enemy, you should cut his cable in several places, so that when he has hopefully mended one break he finds that he still has some more to locate and splice. The neatest trick of all, however, is to cut his cable in such a way that his testing instruments will not show him where the break is.

Very early in the days of cable-laying, techniques were worked out for locating the approximate position of faults, so that the repair ships would not have to grope blindly along the sea-bed for scores or hundreds of miles. In the case of a cable which has a clean break, so that its conductor is short-circuiting into the sea, the problem is a particularly simple one. The electrical resistance of any cable is known when it is laid down and if a section is chopped off, the resistance of the piece that remains will be reduced proportionately. A measurement made at any point along the cable, followed by some simple arithmetic, would show the location of such a break. (Other types of fault, however, may be much more difficult to locate.)

Any reader with a sufficiently antisocial turn of mind will already have worked out how a submarine cable can be sabotaged so that it is impossible to locate the break by straightforward electrical tests. Instead of leaving a clean cut open to the sea, you should attach to the end of the cable a 'dummy-resistance' which is electrically equivalent to the length that has been severed. The tests will then show that the cable is still just as long as it ever was – but it won't carry any signals.

It is one thing to be able to cut your enemy's cables; it would be better still if you could tap them and read his messages. At first sight – especially in these days of telephone-tapping – this would seem a fairly straightforward problem, at least in shallow water. In practice, however, it would be exceedingly difficult for any eavesdropping submarine to interpret the torrent of electrical impulses passing through a modern cable, even if it could pick them up. And having performed so nearly impossible a feat, it would still be necessary to get the information back to base.

I have been able to find no evidence that submarine cables have ever been successfully tapped, and some telegraph engineers have claimed that such a thing is quite impossible. Even if this is not literally true, it is certain that the security of the cable system has never been seriously threatened. In wartime, all secret signals have to go by cable; after a hundred years, the copper wires along the ocean bed remain the safest messengers mankind has yet discovered.

CHAPTER 13

The Deserts of the Deep

Before the coming of submarine telegraphy, nothing whatsoever was known of conditions in the ocean depths. To those who gave any thought to the matter, the deep sea was a place of utter mystery, peopled with hideous monsters, and littered with the wrecks and treasure of centuries. It was as unattainable and remote from human affairs as the far side of the moon.

The picture changed as soon as men attempted to lay the first cables in the open sea, for it then became vitally important to gather knowledge about this unseen realm which covers more than half the world. The telegraph-ships had to know the depth of the water beneath their keels, as well as the type of terrain over which they were sailing, sometimes as high as a cloud above the surface of the earth. Their skippers had to be certain that their cables were not draping themselves over unexpected precipices or mountains; it was also important to know whether the sea-bed was covered with rocks or had any other peculiarity that might affect the operation of the cable, or make it impossible to grapple for it if it developed a fault.

When Lieutenant Maury started collecting material for his *Physical Geography of the Sea*, only 180 soundings had ever been taken in the open Atlantic, beyond the shallow waters of the continental shelf. This was partly because no one was particularly interested, and partly because the lowering and raising of several miles of line with a heavy weight on the bottom was a tedious and time-consuming business. Not until steam winches were available to haul up the lines swiftly and effortlessly did deep-sea soundings become practicable; this is not the first time that a simple mechanical invention has had important and unexpected scientific repercussions.

From 1854 onwards, soundings began to accumulate from the oceans of the world, and methods were devised for collecting specimens from the sea-bed by ingenious grabs and scoops. These techniques have now culminated in machines which can bring up core-samples hundreds of feet long, providing the geologists with millions of years of submarine history.

The invention of these new instruments, the rapid development of deep-sea cables, and the great stimulus given to biological studies by Darwin's *Origin of Species*, all reacted together to produce the first great oceanographic expedition – the classic voyage of HMS *Challenger*. Between 1872 and 1875 this 2,306-ton corvette with her 400 h.p. auxiliary engine travelled right round the world, beinging back a volume of knowledge which has never been surpassed. The expedition was a joint venture of the Royal Society and the British Navy, and though there were only six civilian scientists on board, led by Professor C. Wyville Thomson, they were ably assisted by highly qualified naval officers. The results of their work filled fifty massive volumes, which are still a mine of marine information to this day.

The main result of the *Challenger* expedition was to revolutionise ideas about life in the ocean abyss. Popular imagination might fill the deep with monsters, but the scientists of the early nineteenth century knew better. Nothing could possibly live in total darkness, at a temperature only a few degrees above freezing point – and, worst of all, at a pressure of several tons to the square inch.

The *Challenger* proved that the scientists were wrong. There was life at the greatest depths that the dredges and nets could reach. It was carnivorous, for no vegetation could exist thousands of feet below the reach of the last rays of light, and the only source of food was the incessant rain of biological debris from the ocean's upper levels. On this, and upon each other, preyed legions of nightmare beings – tiny dragons, fish that could swallow creatures several times their own size, phosphorescent squids, sharks with elongated fins which enabled them to rest like tripods on the sea-bed . . .

Such is the alien life that swims and battles above the thin cables which carry men's words and thoughts from land to land. And one thing is certain: even now we have glimpsed only a minute fraction of the menagerie of the abyss. For what would we know of life on the earth's surface if our only information came from dredges let down by helicopters beyond the clouds?

The ocean bed itself is covered deeply with a dense slime or ooze which dries into a hard clay when exposed to air. It is fortunate that this is sufficiently close-packed to support the weight of a submarine cable on its surface, since if cables sank deeply into the ooze, retrieving them for repairs might be impossible.

This deposit is largely composed of the skeletons of myriads of minute creatures known as plankton, which serve the same role in the ocean that plants do on land. They are the beginning of the great food chain which terminates in the higher fish (and often in man himself); when they die their chalky or silicious skeletons, which are miracles of microscopic design, slowly drift to the sea-bed, where they form layers thousands of feet in thickness. Out in the Atlantic basin, indeed, the layers of sediment have been found to be as much as twelve thousand feet thick. Such deposits must have taken not millions but scores of millions of years to accumulate. Their discovery, which is quite a recent one, was a final death-blow to the legend of lost Atlantis. They prove that no continent can have existed in the Atlantic much later than the time of the great reptiles – ages, therefore, before the coming of man.

This endless rain of tiny skeletons, to which must also be added the mud of the continents which the world's great rivers for ever sweep into the sea, has long ago blurred and buried all the minor irregularities of the ocean bed. But the floor of the ocean is not a featureless plain; it is wrinkled with submerged mountains, scarred with trenches and valleys, puckered here and there with mysterious flat-topped plateaux. The mid-Atlantic contains the greatest mountain range on earth, ten thousand miles long and five hundred miles wide, occasionally breaking the surface at such spots as the Azores, where Pico rears seven thousand feet into the sky out of waters twenty thousand feet deep.

It was the northern foothills of this mid-Atlantic range which the surveys of the 1850s revealed, and which Lieutenant Maury christened Telegraph Plateau. Such a label is much too neat (though it was excellent propaganda at the time); the best that can be said about this 'Plateau' is that it contains no exceptionally deep trenches which might be a hazard to cable-laying operations.

Sheer depth is not a serious problem for a well-designed cable, but any abrupt irregularities are an obvious danger, for a cable might be slung across a submarine canyon and subjected to a strain which would eventually snap it. Moreover, in regions where the sea-bed plunges suddenly into great depths there is a strong liability of seismic disturbances. The rocks of the earth's crust are under such enormous strain in such unstable areas that occasionally something gives, and the result is a submarine earthquake.

Such an event caused great alarm in Australia in 1888, when three cables to the continent snapped simultaneously and the country lost contact with the outer world. Not unreasonably, it was assumed that the cables had been cut by an enemy, and the Navy was promptly mobilised to meet the expected crisis.

The danger of earthquakes is one that might have been anticipated; indeed, from the earliest times attempts were made to avoid laying cables through regions where there was liable to be volcanic activity. But a more subtle peril was not discovered until quite recent years, and is still something of a mystery.

On November 18, 1929, a vast submarine convulsion in the North Atlantic snapped most of the cables between Europe and America. But they did not break simultaneously; they went one after another as if some disturbance was moving along the sea-bed. It is believed now that what broke the cables was a 'turbidity current '– a submarine avalanche of silt-laden water triggered off by the earthquake, and travelling initially at about fifty miles an hour. In any event, it took six months to repair the damage, and the loss to the cable companies was over £1,000,000.

Perhaps the most extraordinary accident that has ever happened to a submarine cable occurred near Balboa in April 1932. The repair ship *All America* had been sent to mend a break in water three thousand feet deep, and when – with considerable difficulty – she hauled up the damaged cable the cause of the trouble came with it. A forty-five-foot sperm whale had become hopelessly entangled in the iron coils, which were wrapped around its lower jaw, flippers and flukes.

This was very annoying for the cable company (not to mention the unfortunate whale), but it gave most valuable information about the habits of these great animals. The sperm whale is known to feed on the giant squid, which it hunts in the darkness along the ocean bed, but many naturalists had found it hard to believe that an air-breathing mammal could possibly descend several thousand feet in search of its prey. Yet here was one that had established a diving record of 3,240 feet before it met an enemy it could not conquer and was drowned in the ensuing struggle. Did it mistake the iron cable for the tentacle of a giant squid? It seems possible, though we shall never know for certain. Nor do we know yet to what depths these supreme divers can descend, and how they manage to avoid the physiological problems such as the 'bends' which limit us to the upper few hundred feet of the sea.

To deal with all the accidents which can occur to submarine cables a fleet of repair ships is kept in readiness scattered over the oceans of the world. These ships are quite small vessels – about two thousand tons – as they do not have to carry the heavy loads of their big sisters, the cable-layers. Their job is a skilled, humdrum and frequently unpleasant one, for they may have to operate in most adverse weather conditions.

Today, recovering a damaged cable is no longer the uncertain affair that it was in the days of the *Great Eastern*'s heroic fishing contest. When a break has been reported by the shore stations, and its location found as accurately as possible by electrical measurements, the repair ship proceeds to the spot and puts down a marker buoy so that it has a reference point from which to work. Then it starts dragging operations with a grapnel chosen according to the nature of the sea-bed. If the bottom is sandy, a rigid grapnel is used, with prongs which plough beneath the surface; if the cable lies on rock, the grapnel used is a sort of flexible snake with hooks along its length.

In deep water, where the cable might not be strong enough for lifting in its entirety, a 'cutting and holding' grapnel is employed; this cuts the cable soon after it has been grappled, so that only one end is brought up at a time.

It is possible to tell from the instruments recording the strain on the grapnel-rope whether or not a cable had been hooked. But the officer-in-charge has a more sensitive method of detection; he analyses the vibrations coming up the rope by sitting on it, and many old hands claim that this technique gives much more accurate information than any instrument. There were pioneers of aviation who claimed to be able to fly by the seat of their pants; the men of the cable ships were working on this principle a hundred years ago.

H

Once the ends of the faulty cable have been secured, it is then a routine matter, as long as the weather co-operates, to locate the trouble and splice in a new section. Many of the older cables contain literally hundreds of repairs; sometimes, indeed, about all that can be left of the original cable is its route.

The battle against corrosion, ships' anchors, marine borers, dragging trawls, and even sharp-toothed fish is a never-ending one of which the world knows nothing. Improved materials, as we shall see in the next chapter, have tilted the battle in the cable companies' favour, but anyone who has dealings with the sea must always be prepared for trouble. Sometimes it can be anticipated, but sometimes there are accidents which no one in their senses would ever have imagined. Consider this entry which a disgruntled but somewhat unimaginative operator once inscribed in the log of a telegraph station over-looking the Red Sea – and remember that it refers to the shore-end of a cable, weighing perhaps ten tons in all: 'At 5 minutes past 8 a.m. the cable suddenly disappeared through aperture and has not been seen since.'

What had happened? Well, laying had just started, and the cable ship was less than a mile from shore when the paying-out mechanism jammed. The ship continued on its course – and, despite the strain, the cable did not snap. The whole length, right the way back to the telegraph hut, took off after the ship. One hopes that the engineers, when they had gone back to the beginning and started laying the cable again, had the grace to send a testimonial to its manufacturers.

CHAPTER 14

The Cable's Core

There are two key substances without whose existence the development of submarine cables – and indeed of electrical engineering – would have been impossible. One, copper, has been known since the beginning of civilisation. The other – gutta-percha – was introduced into Europe less than ten years before the laying of the first cable beneath the English Channel.

Copper, either in a relatively pure state or in the form of its alloy bronze, was the first metal man ever learned to work. For thousands of years it was prized for its mechanical properties, yet today those are far less important than its electrical ones. Only silver is a better conductor than copper (by about 10 per cent), and using that for electrical wiring is scarcely an economic proposition. However, it has been done in at least one case where money was literally no object. During the development of the atomic bomb, it was necessary to construct the largest electromagnet ever built in order to separate the isotopes of uranium. The magnet was over a hundred feet across, and providing copper for such a monster would have created a serious drain on the United States' supplies of this vital material. Some genius therefore proposed using the silver which was lying in the Treasury vaults, pointing out that it would be at least as safe inside the closely guarded confines of Oak Ridge. So the United States Treasury handed over fifteen thousand tons of the precious metal to go into the magnet windings; it got over 99·9 per cent of it back when the isotope separator was dismantled and its coils melted down again.*

---

* A. H. Compton, who tells this story in his book *Atomic Quest*, says that the Assistant Secretary of the Treasury was not unduly perturbed at being asked for half a billion dollars' worth of silver, but was distressed when the Army used the phrase 'fifteen thousand tons'. 'Young man,' he reprimanded the colonel who made the request, 'when we talk about silver the term we use is ounces.'

It is fortunate for the communications industry that copper is not yet as expensive as silver; even as it is, the telegraph companies have been engaged for more than a hundred years in a running fight against thieves who specialise in stealing their lines and selling them for scrap. As long ago as 1823 Sir Francis Ronald, whose primitive telegraph system has already been mentioned, clearly envisaged the rise of this parasitic trade and gave his advice for dealing with people who might dig up even buried cables: '. . . render their difficulties greater by cutting the trench deeper; and should they still succeed in breaking the communication by these means, hang them if you can catch them, damn them if you cannot, and mend it immediately in both cases.'

When the first Atlantic cable was constructed, no one realised that the conducting power of copper was greatly influenced by the presence of impurities. The contractors supplied what they considered the best grade of copper, but they were concerned only with its gauge (diameter) and its ductibility or freedom from brittleness. As long as the metal was mechanically good, that was all that mattered. Copper was copper, wasn't it?

Not as far as the electrical or telegraph engineer is concerned. To him, copper with a trace of arsenic or sulphur is no better a conductor than iron. Nowadays, we can go into a radio or hardware store and buy, without giving it a second thought, copper which is purer than anything the early Victorian scientists could make in their laboratories. The wire which carried the first messages across the Atlantic would have been rejected with indignation by any electrical contractor today.

Being able to carry electricity where you want it to go, with a minimum of loss through resistance, is only half the problem. Providing an efficient insulator to stop it from leaking away proved even more difficult in the early days of telegraphy, and it is hard to see how the industry could ever have developed if gutta-percha had not turned up at the exact moment when it was needed.

Strictly speaking, gutta-percha is not an insulant – nothing is – but is merely a very bad conductor. In actual figures, it is a poorer conductor than copper by a factor of some 1,000,000,000,000,000,000,000. This means, to put it in another way, that a square of gutta-percha half a million miles on a side would not let as much electricity pass through it as a piece of copper only one inch square – assuming that the thickness of each sample was the same.

Gutta-percha is a substance much more familiar to our grandparents than it is to us, for it has now been largely replaced by the many synthetic plastics that modern science has produced. The gum of a tree found in the jungles of Malaya, Borneo and Sumatra, it was introduced into Europe in 1843, and its remarkable properties were at once recognised. Indeed, it was the first natural thermo-plastic ever to come into general use. Unlike rubber, it is not elastic, being hard and solid at room temperatures. However, in hot water it becomes as malleable as putty, reverting to its original hardness when cold again. This makes it extremely easy to mould it into any desired shape, and in the 1850s an extraordinary variety of gutta-percha articles came on to the market, such as dolls, ear-trumpets, 'chamber utensils for use in mental homes', pin-cushions, inkstands, chessmen 'not liable to be broken even if thrown violently on the ground' and lifebuoys for ocean voyagers. ('No emigrant ought to be unprovided with them for he can, at the end of his voyage, use the material for shoe-soles.')

Curiously enough, one of the early uses of gutta-percha was in *non*-electric communication over a distance, by means of speaking-tubes. It is impossible to keep a straight face while reading the testimonials and advertisements for these. One would give a good deal the see the Barrett family, on a day's outing from Wimpole Street, employing one of the 'small and cheap Railway Conversation Tubes, which enable parties to converse *with ease and pleasure*, whilst travelling, notwithstanding the noise of the train. This can be done in so soft a *whisper* as not to be overheard even by a fellow-traveller. They are portable, and will coil up so as to be placed inside the hat.' And in the omnibuses – horse-drawn, of course – 'the saving of labour to the lungs of the Conductor is very great, as a message given in a *soft tone of voice* is distinctly heard by the Driver'.*

Somehow, it is difficult to picture a Cockney conductor speaking

* The italics are all due to the copy-writer of the Gutta-Percha Company, who would have been a big man in the advertising world if he had lived today.

in a soft tone of voice, even with benefit of the Gutta-Percha Apparatus. But it was a boon to doctors, one of whom wrote:

> I have had the tubing carried from my front door to my bedroom, for the transmission of communications from my patients in the night. I have brought it to my pillow, and am able with the greatest facility to hold my communication with the messenger in the street, without rising to open the window, and incurring exposure to the night air.

(Ah, that deadly 'night air'! How it terrified our forefathers!)
And what a picture of a bygone age this report conjures up:

> The Gutta-Percha Hearing Apparatus fitted up in Lismore Cathedral, for the use of HIS GRACE THE DUKE OF DEVONSHIRE, has most fully answered the purpose for which it was required. The tubes are conveyed from the pulpit to His Grace's pew (under the flagging, and altogether out of sight), and although their length is between thirty and forty feet, he is able, with their assistance, to hear distinctly every word.

Poor Duke; he must often have cursed the March of Science.

Yet is was a very different form of communication which gutta-percha was to make possible. The great Michael Faraday was the first to realise that this new material might be the answer to the unsolved problem of electrical insulation in the presence of water. Rubber had already been tried, but was found to perish far too swiftly. The first cross-Channel cable, that of 1850, was coated with gutta-percha and nothing else; there was no armouring of any kind, so that it was in fact a wire rather than a cable. All subsequent cables, for the next eighty years, were insulated with the same material or its derivatives; not until the late 1930s did a fundamentally new insulant arrive on the scene – again just when the electricians needed it.

The long reign of gutta-percha was ended by an unexpected laboratory development, which is almost a classic example of the way in which pure scientific research, with no particular thought of a practical application, can produce revolutionary results. For years the submarine cable companies had been trying to improve the electrical qualities of the insulant which Nature had provided, and they had made substantial advances. But in 1933 a group of scientists at Imperial Chemical Industries, working in an entirely different field, produced a substance electrically far superior to anything that is found in the natural world – a substance which has not only had profound effects on communications, but which has also brought many changes in the home.

The ICI scientists took the cheap and common gas ethylene – C_2H_4 – and compressed it under more than a thousand atmospheres. This is a pressure greater than that found at the bottom of the deepest ocean, and the result was startling. The invisible gas turned into a waxy solid, and when the pressure was released *it stayed a solid*. This new substance, which had never existed in the world before, was christened polyethelene – a name which was itself rapidly compressed to polythene. It was produced just in time to provide the thousands of miles of radar and high-frequency insulation used in the second world war. So priceless was it, and so well kept the secret of its manufacture, that at one time the Germans' only source of polythene was crashed Allied bombers. The single small factory which produced the world's entire output thus had the distinction of supplying friend and foe with this wonderful new substance.

Today, polythene is familiar to everyone in the shape of hygienic, unbreakable containers and transparent plastic bags. We make far more things out of it than the Victorians ever did from gutta-percha; and will posterity consider our products just as amusing as great-grandfather's Emigrants' Lifebuoys and Railway Conversation Tubes seem to us?

CHAPTER 15

The Wires Begin to Speak

On the morning of August 4, 1922, the entire telephone system of the United States and Canada closed down for one minute in a farewell tribute to the man who had brought it into existence, and who was at that moment being lowered into his grave on Cape Breton Island, Nova Scotia. Today, the overland radio link of the transatlantic telephone passes the very mountain bearing Alexander Graham Bell's tomb, and it is equally appropriate that the eastern end of the circuit is in his native Scotland.

The telephone was perhaps the last of the simple yet world-shaking inventions that could be made by an amateur working with limited resources. It has sometimes been stated that had Bell understood anything about electricity, he would never have attempted to make such a ridiculous device, as any *real* expert would have known at once that it couldn't possibly work.

This is both untrue and unjust; Bell knew exactly what he was doing, though he was surprised to discover that it could be achieved by such simple means. If we try to forget that we know the answer and put ourselves a hundred years back in the past, every one of us would probably decide that the transmission of speech over long distances would require highly complicated equipment – even if it could be done at all.

For human speech is a most complex phenomenon, by any standards, and fantastically so as compared with the simple dots and dashes of the telegraph code. Graham Bell was more thoroughly aware of this than most men, for he was a professor of elocution, as his father and grandfather had been before him.

When we speak, we launch into the air a rapidly and continually varying pattern of pressure waves. The frequency (rate of vibration) of those waves covers a very wide range. In normal speech, it extends from a lower limit of around fifty cycles a second for a deep bass up to five thousand cycles a second for a high soprano –

a range, in other words, of a hundred to one, or almost seven octaves.

Moreover, in speech we are never dealing with single, pure sounds, such as those obtained from a tuning fork or an unstopped violin string. Dozens of different frequencies exist together at the same time, and their unbelievably complex sum makes up an individual human voice. We can recognise each other by our voices because our ears can detect and analyse all these frequencies, just as by a similar sort of analysis our palates can tell whether we are drinking milk or brandy or beer. If human beings communicated in pure musical notes, like talking tuning forks, we could exchange information as rapidly we we do now – but we should never know to whom we were speaking, if we had to judge by voice alone and had no other sense to aid us.

Any method of speech transmission, therefore, requires that a wide band of frequencies be carried from one spot to another without distortion. Fortunately for the telephone engineer, we can understand each other, and recognise each other's voices, even when the upper and lower frequencies are missing, and the range needed for intelligible speech is thus reduced to the more manageable figure of two hundred to two thousand cycles a second. Only if we need high-fidelity reproduction – which the telephone has certainly never claimed to give – must we worry about the extreme ends of the frequency range.

Though the matter is now only of historic interest, it is worth noting that speech can be sent quite surprising distances by purely mechanical means, without the aid of electricity. We have already mentioned speaking-tubes, which still have limited applications in ships' engine-rooms and elsewhere, but during the 1880s 'wire telephones' of much greater range were introduced in a desperate attempt to evade the Bell patents. These are still sometimes met as children's toys, but are otherwise extinct. They consisted of nothing more than a pair of light diaphragms with a metal wire joining their centres; the speech vibrations were transmitted along the wire, which did not have to be taut or straight and could even be laid on the ground or under water. Ranges of up to three miles were possible, but five hundred yards was nearer the practical limit. At one time attempts were even made to arrange switching systems with such instruments, so that different subscribers could be put in touch with each other; one can only marvel at such misguided ingenuity.

Surprisingly enough, the actual word 'telephone' came into existence before Graham Bell was born. It was used by Professor Wheatstone as early as 1840 to describe a device he had made for conveying musical notes short distances through wooden rods. By the 1870s. dozens of inventors all over the world were trying to achieve the electrical transmission of speech, and it was only a matter of time before someone succeeded. How true that is is proved by the fact that the American Patent Office received Elisha Gray's design for a telephone on the same day as Bell's, but an hour or two later – to the subsequent great profit, needless to say, of the legal profession, which did very well out of the telephone. But Graham Bell was the first man to produce, patent and publicly demonstrate a practical telephone; though others had come very near this, their work had not been published or carried through to a successful conclusion. Bell received the fame, and his rivals are now only footnotes in the history books. There are no second prizes in the race for any great invention or discovery.

Graham Bell was born in Edinburgh in 1847, but when two of his brothers died of tuberculosis and he was threatened with the same disease the family moved to Canada. Bell was then twenty-three and as he was seventy-five when he died one can assume that the cure was successful. He settled first at Brantford, near Toronto, then moved to Boston, where he became Professor of Vocal Physiology – a high-sounding phrase for teacher of voice production and elocution. It was at Boston that the telephone was invented in 1875, and Bell made the basic discovery that led to it while he was working on a quite different project, which he had christened the harmonic telegraph. It is worth looking at this device, because the principle underlying it is employed, in a much more sophisticated manner, in the transatlantic telephone cable and indeed over the whole area of modern telecommunications.

Bell was trying to perfect a method of sending several telegraph messages simultaneously over a single wire. His plan was to use a series of vibrating steel reeds, each tuned to a different musical note, as the sending instruments, and to have reeds tuned to the same set of notes at the receiving end. It was easy enough to convert a simple musical note into an interrupted electric current;

a make-and-break contact on the vibrating reed was sufficient to do that. All the signals would be transmitted together along the line, but each reed at the receiving end would respond only to currents of its particular frequency and would ignore all the others. The messages would thus be sorted out according to their frequency – exactly as we now separate radio stations by tuning from one to the other.

On the afternoon of June 2, 1875, Bell was tuning one of the reeds of the receiver, while his assistant, Thomas A. Watson, in a room about sixty feet away, was looking after the transmitter. The sending reed had stuck, and Watson tried to get it going again by plucking it. He didn't succeed; what had happened was that the make-and-break contacts had become welded together, and a continuous current was flowing instead of the normal interrupted one.

At the same moment that Watson was plucking the recalcitrant reed, Bell had his ear pressed against its opposite number in the other room. He heard, faintly but clearly, the ghostly echo of the twanging spring, and in that instant the telephone was born. Bell realised at once what had happened; though only a single musical note had been transmitted, the principle had been demonstrated. Other frequencies could be transmitted by the same means – including the wide band of them that constitutes speech.

After this, the development of the telephone was largely a matter of working out details. The instrument which Bell finally produced was extremely simple; it consisted essentially of an iron diaphragm placed within the field of a horseshoe magnet. The diaphragm, set vibrating in the field of the magnet by the pressure waves of speech, generated corresponding fluctuations of current which were transmitted along the line. An identical instrument at the other end converted the electric variations back into sound.

The Bell instrument survives virtually unchanged today in all the telephone receivers in the world, and the speakers of our radios are also its lineal descendants. As a transmitter, however, it was inefficient, and it was soon superseded, after a lengthy patent war, by the carbon microphone invented by Edison. This, in slightly modified form, also holds the field to this day, and it is hard to see how either instrument can be drastically altered or improved.

Once invented, the telephone spread over the face of the world with remarkable speed. It was of such universal value and so simple to use (a contemporary advertisement remarked 'Its employment necessitates no skilled labour, no technical education . . .') that there can be very few inventions in history which came into everyday use so swiftly. Within ten years there were well over a hundred thousand telephones in the United States alone; within twenty-five years there were a million, and when Bell was buried thirteen million instruments were silenced.

The adoption of the telephone was, of course, immensely assisted by the fact that the telegraph, using very similar techniques and equipment, had been in use for thirty years, and it was relatively simple to work the telephone over many of the existing lines. If the telephone had been invented first (an improbable but not impossible event) it would have taken much longer for it to have come into general use, if only for the reason that no one would have believed in it. Even as it was, when Bell's representative offered the telephone to the British Post Office in 1877 the Engineer-in-Chief turned it down as 'its possible use was very limited'.

We are not concerned here with the story of the telephone's swift and uninterrupted rise to its present dominating position in social and business life, but a few dates and highlights are worth recording. One now-forgotten episode is the incredible Edison's equally incredible answer to the Bell patent. As already remarked, Edison had an excellent transmitter, even simpler than Bell's; it worked on the principle of variable resistance. The speech vibrations picked up by a diaphragm varied the pressure on a piece of carbon, and the changing pressure produced a resistance – and hence a current – which fluctuated in sympathy with the original speech.

When the Western Union Telegraph Company tried to introduce this brilliantly simple device it at once ran into difficulties. A transmitter was no use without a receiver – and the Bell Company's lawyers were standing in the background waiting to serve the writs if *their* instrument was employed for the purpose. When he was informed of this situation, Edison – who was busy with half a dozen other inventions at the same time – promised to deliver a receiver which would work on entirely different principles from Bell's. He produced it *five days* later; it generated sounds from the friction of a platinum contact against a rotating cylinder of chalk, and the user had to keep turning a handle if he wished to hear what the party at the other end of the line was saying. Naturally enough, this clumsy and complicated device did not last for very long, and when the Edison and Bell interests were merged the great inventor was able to turn his mind to less sterile pursuits than patent-busting.

As an historical curiosity, we might mention that in 1878 Professor Hughes produced a microphone which probably represents the ultimate in simplicity for any scientific instrument. It consisted, believe it or not, of three ordinary nails – and nothing else. Two of the nails were laid side by side, and the third rested across them like the rung of a ladder. When an electric current was passed through this H-shaped arrangement, it became an extremely sensitive detector of sounds or vibrations. Even a fly walking past it could be heard in a telephone receiver connected to the circuit. The minute tremors, causing the points of contact of the nails to move, produced current variations which were turned into audible sound.

Once he had made his great invention, Bell seemed to lose interest in the telephone; probably the prolonged litigation it inspired disenchanted him, as well it might. His fame and fortune were secure, and he spent the rest of his long life experimenting in various branches of science. Among these was aviation, though of somewhat static kind. Anyone who is still a boy at heart will feel a little envious of the fun Bell must have had when, in 1907, he built what is probably the largest kite the world has ever seen. Fifty feet long and twelve feet high, it consisted of twelve thousand small triangular wings which formed a great honeycomb. It was able to lift a man to a height of 150 feet; its passenger was the unfortunate Lieutenant Selfridge, who a few months later had the sad distinction of being the first man to be killed in an aeroplane.

Within a few years of Bell's patent, the world's great cities had become festooned with a cobweb of overhead telephone wires, often installed by bitterly competing companies. War to the soldering iron was not at all uncommon between them, as linesmen destroyed their rivals' circuits in feats of aerial sabotage.

The overhead lines have gone, at least in the cities, but the telephone exchanges remain as the nerve centres without which the telephone itself would be useless. In the early days many of the operators were boys, but this arrangement did not last for long – probably owing to the working of that curious mathematical law with which all efficiency experts are familiar: 'One boy equals one boy: two boys equals half a boy; three boys equals no boy at all.'

Girl operators soon took over entirely; perhaps the telephone did almost as much as the typewriter to emancipate women and to give them independence. It is amusing to read, in the *Pall Mall Gazette* for December 6, 1883, a description of a London exchange which ends as follows:

> The alert dexterity with which at the signal given by the fall of a small lid about the size of a teaspoon the lady hitches on the applicant to the number with which he wishes to talk is pleasant to watch. Here indeed is an occupation to which no 'heavy father' could object; and the result is that a higher class of young woman can be obtained for the secluded career of a telephonist as compared with that which follows the more barmaid-like occupation of a telegraph clerk.

It was obvious that attempts would be made, at the earliest possible date, to link together the telephone systems of Britain and the Continent, and thus to achieve with the new instrument what had been done years before with the telegraph. The first Anglo-French telephone cable was laid in 1891 by the cable ship *Monarch*, whose successor we shall meet later. It was little more than a slightly modified telegraph cable, which was good enough for the job over the relatively short distance involved. But when attempts were made to establish submarine telephone links over greater distances – such as from England to Ireland – the engineers ran into trouble. The problems which had plagued the first submarine telegraphs were reappearing, in a far more severe form. We have already seen how the electrical sluggishness of the early telegraph cables delayed and distorted the signals passing through them. To some extent, this can be overcome by slowing down the rate of working, but obviously no such solution is possible for a circuit which has to carry speech, not Morse code. If you halve the working speed of a telegraph cable, you halve its earning power – but it can still function. A telephone cable which can only transmit speech at half the rate at which a man can talk is, on the other hand, completely useless.

The problem was solved, as far as cables of a few score miles in length were concerned, by the work of Oliver Heaviside, a brilliantly talented but somewhat eccentric mathematical genius whose name is now remembered in quite a different connection. The Heaviside Layer of the upper atmosphere has become familiar to everyone as a result of long-distance radio – though there have been people who spelled it 'Heavyside' and thought it was a description, not a name. Less well known, except to specialists, is Heaviside's remarkable work in mathematics and communications. And still less well known is the man himself; of all the characters who enter this story, he is surely the most baffling, for he belongs to that gallery of English eccentrics of whom Lewis Carroll is the patron saint.

CHAPTER 16

The Man before Einstein

All the events of Oliver Heaviside's quiet life can be summed up in a few paragraphs. He was born on May 18, 1850 – three months before the laying of the first submarine cable – and died seventy-five years later on February 3, 1925. He was almost entirely self-taught, never married, and for much of his life was virtually a recluse, living a hermit-like existence and seeing few visitors.

After working as a telegraph operator in Denmark during his teens, Heaviside returned to his parents' home in his early twenties and never went out into the world again. He produced his most important scientific papers during the 1880s, and his method of working is not one to be recommended. A lover of heat, he would close the doors and windows of his room, light gas fire, oil stove and pipe, and calculate away into the small hours while the temperature rose to the nineties and the oxygen was slowly depleted. Most of his life Heaviside suffered from ill health; in the circumstances it is hardly surprising, and one wonders that he did not suffer the same fate as Emile Zola.

After his parents died in 1896, Heaviside lived completely alone for twelve years. Then he moved to a house in Torquay, Devon, where he spent the remaining seventeen years of his life. For some time he was looked after by a friendly relative – his brother's sister-in-law – but the strain of caring for a genius eventually proved too much for this kind soul, and after eight years she left Oliver to his own devices.

But though Heaviside was undoubtedly a difficult person, quite a few friends penetrated his armour of reserve. From 1919 to the end of his life he was watched over by a local policeman, Constable Henry Brock, who ordered his groceries and whose daughter delivered them to the house. Heaviside expressed his gratitude in voluminous letters, illustrated by many sketches; unfortunately, none of these survived Constable Brock's death in 1947.

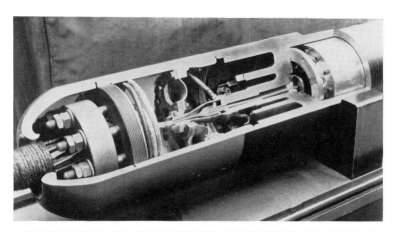

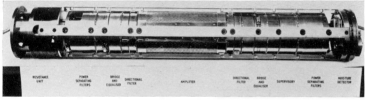

RESISTANCE UNIT	POWER SEPARATING FILTERS	BRIDGE AND EQUALISER	DIRECTIONAL FILTER	AMPLIFIER	DIRECTIONAL FILTER	BRIDGE AND EQUALISER	SUPERVISORY	POWER SEPARATING FILTERS	MOISTURE DETECTOR

Top: 19. Cutaway section of the 1956 Atlantic telephone cable.

Centre: 20. Pressure seal of rigid repeater.

Bottom: 21. Lucas grapnel, for cutting cable before raising it.

Above: 22. Coiling cable in one of *Monarch*'s four tanks.

Below: 23. Foredeck and cable-laying equipment of HMTS *Monarch*.

Though poor, Heaviside was never destitute. Many individuals and organisations did their best to help him, but few succeeded. His early tiffs with conservative mathematicians had embittered a naturally shy and retiring personality, and the fact that he was slightly deaf also tended to cut him off from society. Attempts to aid him financially were sometimes foiled by his stubborn independence; he frequently resembled, to borrow Shaw's description of Mrs. Patrick Campbell, 'a sinking ship firing on its rescuers'.

Heaviside was certainly not a neglected genius; long before his death his great contributions to electromagnetism and telecommunications had been fully realised and he had received that highest of scientific honours, a Fellowship of the Royal Society. (Professor Bjerknes, the great Norwegian meteorologist, once remarked: 'I proposed Heaviside for the Nobel Prize; but, alas, it was a hundred years too early.') The Institution of Electrical Engineers made particularly determined efforts to help and honour him, with fair success. In 1921 it instituted its highest award, the Faraday Medal, and Heaviside was the first recipient. With some trepidation, the President of the IEE called on the old man to present the award; here is his account of his reception:

> Heaviside lived entirely alone in a pleasant house in Torquay – a house decaying from long neglect. I found him waiting in the weed-covered drive in an old dressing-gown, armed with a broom, trying rather vainly to sweep up the fallen leaves. He was pleased to see me in a queer, shy way and took me through a furniture-laden hall, all covered with dust. . . . He vigorously criticised the wasteful expenditure on the leather-covered vellum document which accompanies the medal, but was consoled by the medal being of bronze and not of gold. . . .

One of the few visitors who saw him fairly regularly in later years has recorded the perils of accepting Oliver's hospitality:

> I had first to help him find gas-leaks with a lighted candle. We patched up a flexible leaky gas tube and then he made tea. He put the whole of the contents of a new $\frac{1}{4}$-pound packet of tea into the tea-pot. I had to drink the potion, which he had fortified with a heavy dose of condensed milk. . . . He provided

I

a good cup for my wife and a slop basin for me. Most of his crockery had gone the way of all crockery. A sheet of *The Times* formed the table-cloth. . . .

Though he was an eccentric, Heaviside was certainly no ogre. In his old age he was described as strikingly handsome, with brilliant eyes, a most remarkable head of white hair, and the gracious manners and bearing of a 'gentleman of the old school'.

It is satisfying to record that he ended his days in more comfort than he had lived. Found unconscious by the faithful Constable Brock one January evening, he was taken to hospital (the ambulance was the first automobile he had ever ridden in!) and quickly revived. A great favourite of the nurses, he was full of fun and enjoyed the good food; but his seventy-five years were too much for him and he died four weeks later.

So much for this man; his uneventful life is wholly overshadowed by his work, which appeared in a long series of technical papers and three massive volumes entitled *Electromagnetic Theory*. Many of his results were obtained by a mathematical technique (the Operational Calculus) which caused a minor scandal when he published it, for the purists were unable to prove to their satisfaction that Heaviside was justified in using his equations in the way that he did.

To put it briefly, Heaviside treated mathematical *operators* as if they were *quantities*. The familiar signs of ordinary arithmetic – $\times$, $\div$, $\sqrt{}$, $-$, $+$ are all operators; they have no values in themselves, but are merely orders or instructions. More complex operators are the differential and integral signs met with in calculus, and Heaviside was particularly concerned with the first of these. When such entities occur in equations, they are normally associated with the quantities which they modify, but Heaviside left them up in the air, forming in effect equations which consisted solely of operators which had nothing on which they could operate. This was as bad as writing sentences containing all verbs and no nouns or even pronouns (try it and see how far you get), so it is not surprising that Heaviside's fellow mathematicians were up in arms. But the method worked – usually; though as Sir Harold Jeffreys has remarked: 'Heaviside got many wrong answers, but by amazing ingenuity and industry in calculation he was able to find his mistakes. The fact that he succeeded, however, is no guarantee that everyone else could do so. . . .'

Such unorthodox techniques did not make it easy to follow Heaviside's mental processes, and to one scientist who protested that his papers were very difficult to read he made the now classic retort: 'That may well be but they were much more difficult to write.'

In his researches into the very foundations of physics, Heaviside became aware that mass and energy were equivalent long before this was generally realised by the scientific world. By 1890 he had already arrived at a rigorous proof of the famous relationship $E = mc^2$, thus anticipating Einstein's more general formulation of this law by some fifteen years. This is perhaps his most astonishing – and least known – achievement.

Also like Einstein, Heaviside spent the last years of his life working on a Unified Field Theory that should weld together electricity, magnetism and gravitation. He had incorporated his results in Volume 4 of his *Electromagnetic Theory* – but they were never published, and despite extensive searches the manuscript of this volume has never been found. It is known to have existed, for Heaviside submitted it to an American publisher, who understandably baulked at the £1,000 advance he demanded.

And here is a tantalising enigma that may never be solved – like the mystery of Einstein's dying words, which escaped into the unknown because the nurse at his bedside knew no German. There must, surely, have been a copy of the manuscript in Heaviside's house when he was taken to hospital, but no one thought of looking for it at the time. Unfortunately, when the announcement of Heaviside's death was broadcast by the BBC, an enterprising burglar broke into the empty house. He could not have found much of value (and how bitterly Constable Brock must have regretted being unable to perform one last duty for his old friend!), but many books and papers were stolen and scattered. Perhaps one of the keys for which the world's physicists have searched in vain for a generation was lost on that February night in 1925.

However this may be, Heaviside left enough behind him to secure his place in mathematics and, above all, communications theory. As Lord Kelvin had done thirty years earlier, he tackled the problem of the current flowing in a long submarine cable, but he was concerned now with the complex and high-speed impulses of speech, not the relatively slow ones of telegraphy. To work satisfactorily, a telegraph cable must be able to transmit between one and two hundred impulses a second, and a certain amount of distortion can be tolerated, since a Morse signal can be reshaped or regenerated by suitable receiving equipment and made as good as new. To transmit speech, however, at least 2,500 impulses a second must be handled, with no appreciable distortion. The low frequencies of the male voice at its gruffest, the high frequencies of an indignant soprano – all must travel along the line with equal facility.

Needless to say, in general they do nothing of the sort, and there are two effects which make it impossible to send speech for any great distance through a simple submarine cable. The first is attenuation, or the inevitable fading out of the signals as they pass along the line. To make matters worse, the higher frequencies fade out more rapidly than the lower ones – an effect which also occurs with sounds in everyday life. If you hear a brass band a long way off (which is where many people prefer it) all that you can make out at first is, in Omar's phrase, 'the sweet music of a distant drum'. It is not until the band comes nearer that you can distinguish the higher-pitched instruments such as the fifes. Even in the air, the low frequencies carry better, and this effect is much exaggerated in a submarine cable.

To some extent, this tendency could be counteracted by 'boosting' the higher frequencies, thus making up for their increased losses. This is what we do when we turn up the 'treble' or 'top' control of a hi-fi set, in an attempt to correct the characteristics of a recording or the deficiencies of a loud-speaker. However, there eventually comes a point when there is nothing left to amplify, and no amount of boosting is then of any use.

A subtler, and even more serious, form of distortion is caused by the fact that the different frequencies also travel at different speeds through a cable. This effect, luckily, does not happen with sounds propagated through the air. If it did, the results would be extremely odd. Music would be impossible; at a symphony concert, if all the instruments *simultaneously* sounded a note at the middle of their registers, the audience would hear the piccolos first, then the violins, then the 'cellos, and the double basses and contra-bassoons last of all. Even speech would be impossible, unless we agreed to converse with each other at a constant distance. If I spoke the word 'Nonsense', by the time it reached you the 'sss' at the end would have overtaken the lower-pitched 'nnn' at the beginning, and the word would have turned into the thing it described.

These peculiar effects are caused almost entirely by the excessive electrical capacity of submarine cables, which we have already mentioned in Chapter 5, where we compared a cable to a hosepipe which takes a definite length of time to fill up, so that one has to wait before anything comes out of the far end.

However, a cable also possesses another electrical characteristic, known as inductance.* The mechanical equivalent of this is inertia; an electrical circuit, like a piece of matter, has a certain sluggishness and takes some time to respond when an impulse is applied to it.

A submarine cable has very little inductance, and at first sight this might seem to be a good thing. However, when he had completed his mathematical analysis Heaviside discovered, no doubt to his surprise, that if one deliberately *increased* the inductance of a cable its transmission characteristics would be improved. What happens cannot be explained in non-mathematical terms, but we may say that a cable's inductance and its capacity tend to counteract each other. By correct adjustment, indeed, they can cancel out completely, and the result is what Heaviside called a 'distortionless line' – that is, one in which *all* frequencies travel at the same speed and suffer the same attenuation or fading.

It was ten years or more before the engineers appreciated and accepted this peculiar result; perhaps they were as suspicious of Heaviside's equations as the pure mathematicians were, though for different reasons. But eventually it was proved by experiment that submarine cables could be greatly improved by deliberately adding inductance to them, either by inserting coils at regular

---

* The two other cable characteristics, resistance and (ugh!) leakance we will lightly ignore. They are negligible compared with capacity and inductance.

intervals along their length, or by winding iron wire around the central conductor.

This discovery of Heaviside's, put into practice by Michael Pupin in America and Krarup in Denmark (Heaviside being then a prophet without much honour in his own country), made submarine telephony possible over distances of a few hundred miles. Inductive loading, as it was called, was also applied to telegraph cables, greatly increasing the traffic they could carry. The latest Atlantic telegraph cable works at five times the speed of an unloaded one, and therefore has five times the revenue-earning power. Even before his death, Heaviside's equations were earning thousands of pounds a day. There is big money in mathematics, but seldom for mathematicians.

By the late 1920s, improved insulating materials and special alloys for inductive loading had made it possible to think seriously of a telephone cable across the Atlantic. The pioneer in this field was Dr. E. O. Buckley of the Bell Telephone Laboratories; between 1928 and 1931, in conjunction with the British Post Office, he carried out a series of experiments with sample cables off the coast of Ireland and in the Bay of Biscay. Unfortunately, a single cable could carry only a single conversation for such a distance and this made the project uneconomic. To improve the performance, the use of amplifiers in the cable was considered; these were first visualised as sunken globes, anchored to the sea-bed, and carrying batteries for six months' operation.

Here was the germ of the idea which was to lead, a generation later, to the submerged repeaters of today's Atlantic telephone cable. But nothing came of the scheme at the time, for two main reasons. The first was the economic uncertainty of the 1930s, which made it unlikely that such a technical gamble would pay; the other was the development of radio, which provided an entirely new and unexpected method of long-distance communication, besides giving the submarine cables the greatest challenge of their career.

At this point, therefore, we have to make a wide detour into a field which would have seemed as miraculous to the pioneers of Atlantic telegraphy as their enterprise appeared to many of their contemporaries. The human voice spanned the Atlantic by radio forty years before it made the same journey by cable, and the submarine telephone system could never have been built without the use of many techniques developed for radio.

So the detour is necessary – and the change of scenery may be stimulating as we move from the cold, dark depths of the ocean to the seething electric cauldron of the ionosphere.

CHAPTER 17

Mirror in the Sky

The existence of radio waves was first discovered by the great mathematical physicist James Clerk Maxwell, sitting in his Cambridge study and writing equations. He proved theoretically that when an electric current oscillates in a conductor, it throws off waves which travel through space at the speed of light, and which in fact differ from light merely by possessing much longer wavelengths and hence lower rates of vibration.

Maxwell did not live to see his equations triumphantly verified. He died in 1879 at the early age of forty-eight; eight years later, in a series of classic experiments, a young German scientist named Heinrich Hertz became the first man to generate and detect the waves which were to revolutionise communications and to change patterns of culture and society over all the world.

Ironically enough, Hertz did not believe that his work – important though it was to the understanding of the physical universe – would have any practical consequences, and specifically pooh-poohed the idea that radio waves could ever be used for signalling purposes. This kind of blindness to the results of their own work is not uncommon among physicists (as well as other people). Lord Rutherford, the first man to split the atom and unravel its structure, used to laugh at imaginative journalists who wanted to know if atomic energy would ever be harnessed. 'We'll always have to put more energy into the atom than we'll ever get out of it,' he stated categorically – and missed refutation by Hiroshima by exactly the same number of years that Maxwell missed confirmation by Hertz.

It is seldom that a single man dominates an important and rapidly expanding field of technology, but for thirty years Marconi was the Colossus of radio. He was scarcely out of his teens when he succeeded in sending radio waves for a distance of a mile near Bologna, Italy, and two years later – in 1896 – he moved to England, where many of his most famous experiments were carried out, frequently in connection with the British Post Office.

Very early in the development of the art, it was discovered that radio transmitting and receiving equipment could be tuned, so that one could choose the station one wished to listen to, and ignore all others. We take this so much for granted that it is hard to realise that someone had to discover it; the credit is due to Sir Oliver Lodge, who first demonstrated the principle in 1897.

As the twentieth century dawned, radio (or wireless, as most people then called it) rapidly extended its range, and in 1901 it leaped the Atlantic. Flying a receiving antenna from a kite in Newfoundland, Marconi was able to pick up Morse signals transmitted from Poldhu, Cornwall.

Here was a first-class mystery. *If* radio waves behaved like light, there was no way in which they could bend round the curve of the earth. A searchlight in Cornwall, no matter how powerful it was, could not be seen more than a few score miles out in the Atlantic; after that distance its rays would have arrowed on out into space, high above the falling curve of the world.

In 1902 Oliver Heaviside (and, simultaneously, Kennelly in the United States) proposed an explanation which seemed almost as far-fetched as the facts. They suggested that, at a very great altitude in the atmosphere, there was a reflecting layer which turned radio waves back to earth, so that they did not escape into space. As it seemed most unlikely that Nature should be so considerate to the communications industry, and it was also hard to see what could create a layer with such peculiar properties, scientists were slow in accepting this explanation. Not until 1924 – only two months before Heaviside's death – did Appleton and Barnett prove conclusively that the upper atmosphere contained not only one reflecting layer but at least two. Today hundreds of rockets – and dozens of astronauts – have flown through the ionosphere, and many of its secrets have been uncovered.

The early radio workers had been hampered by two serious deficiencies in their equipment: their methods of detecting the waves were very insensitive and cumbersome, and they had no way of amplifying the signals when they had been received. Radio, in fact, was still in the pre-crystal-set stage.

The first major break-through came in 1904 when Fleming

invented the diode valve, the primitive ancestor of all the billions of electron tubes in the world today. The name 'valve' was accurate enough; the diode allowed signals to pass in one direction, but not in the other. It turned the rapidly varying radio waves into audible signals – but it could not amplify them.

That essential step came in 1907, with de Forest's invention of the triode. By feeding the faint impulses to a wire-mesh grid strategically placed inside Fleming's diode, de Forest made the overwhelmingly important discovery that it was possible to amplify signals to an almost unlimited extent. The triode ushered in the electronic age, in whose first dawn we are now living, and was therefore one of the truly epoch-making inventions of history.

In the field of communications, where it received its first use, the triode and its more complex successors gave radio the basic tool needed for its swift development. Once the means for amplifying faint and rapidly varying electric currents had been discovered, armies of ingenious engineers, with Marconi well in the forefront, worked out the rest of the radio technology and built up the most swiftly expanding industry the world has ever seen.

The early experimenters, once they had got over their surprise at discovering that radio waves could bend round the earth, quickly investigated the laws controlling their propagation. They found that the longer the wave, the greater the range at which it could be picked up; for his transatlantic experiments, Marconi used waves almost a mile in length. These long waves needed correspondingly huge antenna systems for their radiation and reception, and a long-wave radio station is a most impressive sight, with arrays of towers hundreds of feet high and covering square miles of ground. Until the 1920s, these immense installations appeared to be the only means of establishing round-the-world radio circuits. The short waves, being of no use except for local communication, were handed over to the amateur experimenters or 'hams', who accepted them grudgingly, protesting at the injustice of their treatment. They did not know it, but they were rather like Oklahoma Indians being fobbed off with a piece of unwanted desert that just happened to be soaked with oil.

In the early 1920s, the ham operators made a discovery which brought the governments and communications firms back into the short-wave field in a hurry. The early tests on these waves had shown that their range was very limited, and also somewhat variable; they faded out a few scores of miles from the transmitter. What no one had dreamed was that they came in again, often loud and clear, thousands of miles away, after being reflected from the ionosphere.

It is not surprising that it took some time to discover this. After all, if one was carrying out tests between, say, New York and Washington, one would hardly bother to place additional receivers in Greenland and Peru on the off chance that signals could be picked up there. Not until the world was well covered with enthusiastic amateurs busily searching the radio spectrum and trying to beat each other's distance records did the unexpected pattern of short-wave reception come to light.

In 1924 Marconi, with great technical and commercial courage, decided to exploit the possibilities of the short waves. At that time the world's long-distance radio links employed waves of five to ten *miles* in length, generated at very high power levels and broadcast from huge and expensive antenna systems. Marconi believed that much better results could be obtained far more cheaply by using waves yards, not miles, in length.

The rest of the world was sceptical; though short waves could be received over vast distances, reception was erratic and apparently unpredictable. Marconi hoped to overcome this by using beam systems, so that most of the radio energy would be sent in the desired direction and not broadcast wastefully over the whole of space. Only with the relatively small antenna arrays which the use of short waves made possible could this be done economically; attempts to make directional antennae on the long waves had resulted in systems up to ten miles in length, and of poor efficiency at that.

Marconi's gamble was brilliantly successful, and during the period 1927–8 Britain was linked by short waves with Canada, India, South Africa and Australia. The new radio service was so efficient, in fact, that it was a serious threat to the existing submarine cables. In 1928, therefore, the British cable and radio interests were merged into one body – now known as Cable and Wireless Ltd. – which for half a century has dominated international communications. C. & W. is a typically British compromise between private industry and Government control. The Government is represented on the board of the company, and has a right to take it over in time of war. It is a considerable tribute to the company that this right was not exercised in 1939–45.

To the general public, the word 'radio' is almost identified with the broadcasting of speech and music – radio-telephony, in other words. But radio had existed for a generation before sound broadcasting came into prominence, and even today most of the world's commercial, as opposed to entertainment, circuits handle telegraphic messages, not speech. If one could stand out in space and watch the pattern of radio waves that now covers the Earth, perhaps its most prominent feature would be the narrowly defined beams leaping from ground to ionosphere and back again, ricocheting round the globe as they carry hundreds of words a minute from country to country. By comparison, most broadcast and television stations would be no more than local, though frequently very intense, patches of illumination.

We have already mentioned Marconi's spanning of the Atlantic in 1901, when the letter S (dot dot dot) was transmitted from Cornwall to Newfoundland. It was not until 1915 that the human voice made the same journey, this time in the opposite direction. After a long series of experiments with the transmitter of the United States Naval Wireless Station at Arlington, the American Telephone and Telegraph Company picked up intelligible speech *via* a receiver at the top of the Eiffel Tower.

The experiments were carried out under difficulties, for the Eiffel Tower was the centre of the French military communications system and the antenna could be spared only for a ten-minute interval in the small hours of the morning. After several months of patient waiting and adjustment of the apparatus, occasional words were picked up, and the first complete sentence was received at 5.37 a.m. on October 23, 1915. For the record, the spoken words which blazed a trail for some many millions of others across the Atlantic were 'Hello, Shreeve! How is the weather this morning?'

The first commercial radio-telephone service between New York and London opened in February 1927, using a wave-length of some 6,000 metres – about four miles. This was sixty-one years after the establishment of the submarine telegraph, and fifty-one years after the invention of the telephone. From that date until the laying of the submarine telephone cable in 1956, radio was the only means of sending speech across the Atlantic.

Unfortunately, it was not a wholly reliable means. Though great improvements were made in transmitters and receivers, nothing could be done about the third link in the chain – the ionosphere itself. When conditions were good, transatlantic speech was of excellent quality, with little distortion or interference. But all too often the radio beams picked up most peculiar noises, like the sounds of cosmic frying-pans. These were usually no more than annoying, but sometimes they could obliterate the signal. There might be periods of hours, or even days, when radio-telephony was quite impossible, and the resulting delays were both infuriating and expensive to the customers.* The Atlantic telephone service was in much the same position as the early airlines; it could never guarantee to be working at any particular time – it all depended on the weather. The weather in this case, however, was not something that concerned the first few miles of the atmosphere, but the last few hundred.

The study of the ionosphere is one of the most intricate branches of modern science, as well as one of the most important both from the practical viewpoint and from the light it throws on the universe around us. To look at it in any detail would take us far outside even the generous limits of divagation set for this book, yet it is necessary to say something about the ionosphere's causes and idiosyncrasies to understand why, after a thirty-year battle, the telephone engineers turned at last from the upper atmosphere back to the depths of the sea.

The ionosphere is neither a simple nor a stable structure; it consists of three main layers, the lowest (E layer) about eighty miles up, the higher (F_1 and F_2) layers ranging round the 150–200-mile level. The designations E and F, incidentally, were given by Appleton, who was the first man to discover that there was

---

* To forestall indignant protests from the Post Office and the other bodies concerned, I am well aware that this description is unfair to the short-wave service, and that the vast majority of calls were put through with little delay. But the fact remains that the only time *I* ever tried to radio-telephone New York from London it took me two days to get through.

more than one layer. With laudable foresight he started at the letter E in case any further layers turned up nearer to the ground – as indeed they have.

We know now that the chief agency producing these layers is the flood of ultra-violet light falling upon the Earth's atmosphere from the sun. Ultra-violet light is generally regarded as being health-giving, and so it is – in weak and feeble doses. The raw rays from the sun, however, would destroy all terrestrial life within minutes if they reached the surface of the Earth; fortunately for us, they are filtered out miles above our heads. And as a by-product of this filtering process they electrify (ionise) the atmosphere, spending their energy in tearing electrons from the widely spaced atoms of oxygen and nitrogen they encounter. Air which is strongly enough electrified reflects (or, more accurately, refracts) radio waves, just as air under suitable conditions of temperature reflects light waves and thus produces mirages.

Since the ionosphere is maintained by sunlight,* it naturally changes in density and altitude between day and night, summer and winter. It is possible to allow for this effect to a considerable extent, by varying the wavelength employed, but there are limits beyond which no such technical tricks are of any avail.

As in the lower atmosphere, the sun is both the creator and the disturber of the weather. It maintains the ionosphere, but sometimes it tears it to pieces by blasts of intense ultra-violet radiation emerging from violent explosions on the solar surface. Some of these are associated with sun-spots, which vary in frequency over an eleven-year cycle, so that at one period the face of the sun may be freckled with dark whirlpools many times the size of the earth, while at another it may be completely unmarked. It is at the times of peak activity that the ionosphere is most disturbed, and radio communication correspondingly upset.

We may think of the ionosphere, therefore, as an earth-englobing mirror which pulsates with the days and the seasons, which is seldom smooth or perfectly reflecting, and which is sometimes shattered into fragments which may take hours or days to re-form. Such a mirror would not be very satisfactory for ordinary use, and

---

* Though not entirely, as meteor dust makes an important contribution.

it is rather surprising that the radio engineers have been able to take as much advantage of it as they have.

But before we abandon the stormy heights of the ionosphere and turn again to the calm stillness of the ocean bed, let us recall one immeasurable debt which civilisation owes to the scientists who probed these electrified layers on the frontiers of space. The radio-pulse technique they used became, a decade after its first employment by Breit and Tuve in the United States, the weapon which won the Battle of Britain and, later, the Battle of the Atlantic. Without radar, pioneered by Sir Robert Watson Watt and a handful of co-workers in the late 1930s, the Luftwaffe would have destroyed the far smaller Royal Air Force, the invasion of Britain would have gone ahead, and we today would be living in a very different world.

Compared with radar, such developments as rockets, jet propulsion and even atomic energy had little effect upon the progress or outcome of the second world war. And radar evolved directly from the pulse-and-echo method of sounding the ionosphere – that remote and invisible layer whose very existence was still unsuspected only a lifetime ago.

There are still foolish people who insist on asking the use of pure scientific research. Nothing could have seemed more detached from everyday life than attempts to measure the electron density a hundred miles up in the air. Yet from this work came the decisive weapon which won the greatest of wars and changed the course of history.

CHAPTER 18

Transatlantic Telephone

At this point I am reminded of the Scots preacher who remarked to his congregation during the course of a sermon on the Holy Writ: 'Now we come to a verra difficult passage, and having looked it squarely in the face, we pass on.' Unfortunately, I cannot take such an easy way out; a good deal of what follows will be 'verra difficult', but it is the reader who must decide whether or not to pass on.

From the nature of things, an engineering project like the Atlantic telephone cable involves matters of great technical complexity, which can be fully appreciated only by those with a good knowledge of electronics. Nevertheless, I believe that all the fundamental problems and their solutions can be understood even by anyone who is terrified at the thought of changing a burned-out lamp-bulb.

I have, therefore, adopted a two-stage approach to the subject. This chapter is entirely non-technical (at least, it is intended to be). A few sections may require reading a couple of times, but I am optimistic enough to hope that at least the general idea will get across to anyone who perseveres to the end.

At the same time, there are thousands – indeed, millions – who have a nodding acquaintance with electronics and would like to have a few more details of the way in which some of the specific problems of transatlantic telephony were solved. I hope that Chapter 21 will make them happy; it is intended for the hi-fi fans, the legions who came into contact with radio or radar during the war, and, of course, all those to whom electronics is their everyday bread and butter.

We have seen how the telephone spread swiftly over the world within a few years of Graham Bell's invention in 1875. But long-distance telephony – even on land – was not practical until forty years later, when the problem of amplifying speech currents was solved by de Forest's triode tube. It had been a fairly easy matter

Above: 24. The author standing beside the Intelsat 4 communications satellite. Tacsat on left—Syncom, Early Bird (Intelsat 1) and Intelsat 2 in background.

Below: 25. Early Bird (Intelsat 1) undergoing radio tests before launch.

26. Early Bird (Intelsat 1) undergoing spin tests before launching.

to boost the fading pulses along a telegraph line by means of the relay or repeater, but doing the same thing for the telephone had baffled the best brains in the business for decades.

Today, when you make a long-distance call, your voice is amplified by banks of electronic tubes at repeater stations forty or fifty miles apart; without such amplification, it would have faded below hearing after a few hundred miles. However, that is the least remarkable of the adventures your voice undergoes; what is not generally realised is that long-distance telephony (and telegraphy too) now relies on what are essentially radio techniques, even though copper wires form the transmitting medium.

From the earliest days of the art, one of the goals of the communications engineer has been to send the maximum number of messages over a single line. We have seen how Bell's telephone was invented as a by-product of his 'harmonic telegraph', an attempt to operate half a dozen telegraphs on one wire by using tuned reeds which would sort out the different messages according to their frequency or pitch. This, of course, is no more than a simple application, in the range of audible sounds, of the tuning principle which enables us to select whichever radio station we wish from the Babel of the crowded ether.

The same principle is now used to send scores, hundreds or even thousands of telephone conversations over the same conductor, by what is known as carrier-frequency transmission. All this means is that when you speak into your telephone, the varying electric currents produced are not themselves transmitted to your hearer. Instead, at the telephone exchange, they are used to control or modulate what is in effect a tiny radio transmitter – one of thousands, arranged in endless racks. The output of these transmitters, each of which is tuned to a slightly different frequency, is what travels over the long-distance lines. The multitude of different conversations do not interfere with each other any more than do all the different programmes surging down the lead-in of your radio or TV antenna.

At the receiving end, the different messages are separated by tuning devices (known as filters, a term which describes their operation rather aptly) and are turned back into audible speech. Because the quality of telephone transmission is a good deal poorer than that which we will accept from the radio, it is possible

K

to squeeze about twice as many telephone circuits as one can pack radio channels into a given band of frequencies. Otherwise, there is no fundamental difference between the two techniques.

The process may be carried one stage farther. It is possible to use a single telephone channel to carry up to twenty-four simultaneous telegraph circuits, tuned even more closely together. Nowadays, in fact, the old distinction between telephone and telegraph circuits has largely vanished. Both services go over the same lines and are handled by essentially similar techniques, and the hundreds of separate wires which formed the older type of telephone cable have been replaced by a single conductor.*

That conductor, however, is something that would have puzzled Graham Bell, Edison and the other pioneers. The old-fashioned pairs of insulated wire, like miniature electric flex, have now been replaced by the coaxial cable, in which a central wire is surrounded by a hollow tube, the two being kept apart by a sleeve of (usually) polythene. Since the advent of TV, the coaxial cable has invaded most of the homes in the United States and Great Britain, but it was originally developed for multi-channel telegraph and telephone working.

The coaxial cable (before long the noun will be jettisoned, and the adjective has already contracted to 'co-ax') can carry an enormous range of frequencies. The lead-in to your TV set is carrying currents oscillating at the rate of at least fifty million cycles a second – and a great deal more, if you are in a VHF area. It would have no difficulty in managing several billion cycles a second if the need arose. In other words, it could carry – at least for a short distance – a million separate telephone conversations without mutual interference.

The three basic elements in modern long-distance telephony, therefore, are the coaxial cable which provides the physical link, the repeater stations every forty miles or so which boost the signals to make up for the losses in the line, and the terminal equipment which merges (at the sending end) and sorts out (at the receiving end) the hundreds of messages passing along the single copper

---

* Cables containing a thousand or more pairs of wires still have to be used in urban areas where there are large numbers of subscribers very close together; the carrier-frequency systems are essentially for long-distance work.

wire inside its hollow tube. With properly designed repeaters, spaced at the right intervals, there is no practical limit to the distance over which telephone conversations can take place. Certainly distance is no physical limitation (though it may be an economic one) to telephony on the continents of this rather small planet of ours. Indeed, it is most unlikely that any planets can exist, anywhere in the cosmos, whose antipodes could not be put in touch with each other by virtually the same equipment which is used when New York talks to San Francisco, or London talks to Rome.

Since the second world war, and as a direct result of developments in radar, the coaxial cable has been challenged by a major rival – the microwave link. Most people are familiar with the tall towers, crowned with enigmatic horns, parabolic reflectors, or funnels, which now rear from the roofs of telephone exchanges or stand in lonely isolation on remote hilltops miles from anywhere. Once again, these are amplifier or repeater stations, but they are connected together not by a copper wire but by narrow beams of radio waves. These beams are so sharply focused that, could they be observed by the naked eye, they would look like searchlights. The towers must, therefore, be close enough to 'see' each other, which is why they are situated on the highest possible ground. Usually they are about as far apart as the repeaters in the coaxial cables – say forty miles – but much greater separations are possible in mountain areas.

The great advantage of the microwave link is that it can leap effortlessly across country through which it would be very difficult and expensive to bury a cable. The obstacles to land lines, incidentally, are not always geographical. Hard-fisted farmers on the make can be as big a nuisance as marshes, rivers and ravines, but the microwave beam ignores them all.

Whether we use coaxial cables or microwave towers, therefore, the maximum length of a single link in a telephone chain carrying a large number of simultaneous conversations is about forty miles. After this distance, the signals need amplifying again if good-quality speech is to be transmitted. This does not matter on land, but it means that any large expanse of water is an apparently insuperable barrier to telephone circuits.

By using specially designed cables, this forty-mile limit can be extended somewhat. In 1947, for example, an eighty-five-mile-long submarine cable was opened between England and Holland and it could carry eighty-four simultaneous telephone conversations. Today, there would be no particular difficulty in designing a cable which could carry a smaller number of conversations for, say, two hundred miles in a single jump.

But two hundred miles is only a tenth of the distance across the Atlantic. Very well, someone may say, can't the problem be solved by making the signals ten times more powerful at the sending end, or increasing the amplification ten times at the receiving end?

Unfortunately, this arithmetic is highly fallacious. The current in a submarine cable does not diminish merely in proportion to distance, but at a far more rapid rate – according to a kind of high-powered compound-interest law, in fact. As a result, one soon runs into some quite staggering numbers (page 165) which make even the figures encountered in astronomy look very small indeed.

No; brute force is not the answer. It has been calculated that if the output from *all* the power stations on earth could be poured into the present Atlantic telephone cable, after only two hundred miles, or one-tenth of the way across, this bolt of energy would be so attenuated that it could not be detected even by the most sensitive of instruments. If this seems a paradox, in view of the fact that a battery the size of a thimble can send an easily read telegraph signal through a transatlantic cable, the answer lies in the frequencies involved. When one uses waves oscillating hundreds of thousands of times a second, the losses are enormously greater than with simple direct currents.

In any event, there is a limit to the amount of power that can be fed into a cable without destroying its insulation or melting its conductors. The demise of the 1858 cable under the treatment meted out to it by Dr. Whitehouse's giant spark coils is a sufficient reminder of this.

Nor can one increase amplification indefinitely; after a while the only effect is to produce noise. If you tune your radio receiver to a blank spot between stations, and turn the gain full up, you will hear a steady hissing sound. Submerged in that ceaseless sibilation will be the programmes from countless transmitters, but you can no more detect them than you can hear the words you spoke a minute ago. The reason is the same in both cases: the signal has sunk below the general level of noise – the random omnipresent agitation produced by all the atoms in the universe.

In a radio set, or any kind of amplifier, most of that noise is due to the fact that the flow of current is not really smooth. An electric current, when examined closely enough, is not like a river, but more like a stream of sand. The individual electrons produce minute random fluctuations, and it is these, when amplified, that mask faint signals and prevent them from being heard.

Once these points are understood, we can see how the problem of transatlantic telephony had to be solved. The only way a multi-channel circuit could be established was to use what were essentially the techniques employed on land – that is, to have amplifiers at short intervals along the route, which boost the signals before they became so faint that they were lost in the background noise which exists in all electrical conductors.

It is simple enough to say this, but the practical difficulties were so great that for a long time there seemed no hope of overcoming them. The average telephone repeater or amplifier is the size of several large filing cabinets and has to have an absolutely steady and reliable power supply. Although it operates for long periods with little attention, adjustments have to be made from time to time, and occasionally components wear out and have to be replaced. In particular, the amplifying tubes are the weak points. As all owners of radios and TV sets know, they slowly – and sometimes not so slowly – deteriorate. This would not be so bad if the ageing could be predicted, but even in the absence of accidents a radio tube's expectation of life is like that of a human being. It may die tomorrow; or it may last for another fifty years. No one can tell.

Designing telephone repeaters which would function faultlessly for decades at the bottom of the Atlantic, under pressures of tons to the square inch, must have appeared such a formidable problem that any reasonable alternative would have been accepted. There are in fact three, and it is worth looking at them if only to see why they were turned down.

It is probably safe to say that the first was never considered. A telephone circuit from Europe to America could be built almost entirely overland – if it went across the USSR. The only submerged section would be under the Bering Straits, and as the distance involved here is about a hundred miles it could be spanned quite easily with a single length of cable.

As we have seen in Chapter 11, this route was attempted in 1864, after the failure of the first Atlantic telegraph. One argument raised against it in the 1861 inquiry is still all too valid today, almost a century later: 'The principal objection would be the internal regulations, and the political character of Russia.'

Even in a sane world, however, it seems unlikely that such a roundabout route would be an economic proposition. Much of the terrain to be crossed would be appalling, and several hundred repeater stations would have to be built and maintained along an arc spanning a large fraction of the Earth's circumference. So whether we like it or not, we have to tackle the Atlantic crossing.

What about using microwave links, from towers on ships anchored at forty-mile intervals across the ocean? We would need fifty of them, with all their crews and equipment. The capital investment would be bad enough, but the running costs would be even worse. There are also some slight practical difficulties; it is very hard to picture fifty ships maintaining their exact positions from Europe to America during a typical Atlantic gale and – this of course is equally vital – keeping their beams precisely lined up on each other despite all their pitching and tossing. They would also prove rather unpopular with other ships, and every so often one of them would be involved in a collision or other maritime disaster and the whole system would be put out of action until it was replaced.

A much more attractive answer would be to use aircraft flying at great altitudes. A plane at forty thousand feet can establish line-of-sight radio contact with another at the same altitude five hundred miles away, around the curve of the earth. (This is why the United States has recently developed airborne radar stations to give increased warning time.) Only four planes would be needed to span the Atlantic, which is rather better than fifty ships.

Even so, capital and running costs would have been enormous, since stand-by aircraft (and crews) would have been needed, and the operational problems would have been severe. Yet it *could* have been done, had there been no alternatives; and this is essentially the solution which, only a few years later, the communications satellites were able to provide in a much more elegant way. In the 1950s, however, the only practical answer was the submerged repeater on the bed of the Atlantic – or, rather, 102 of them, strung between the continents like beads on a necklace.

And that is not a bad analogy, since no product of any jeweller was more valuable, or built with greater care and ingenuity.

CHAPTER 19

The Dream Factory

It would be interesting to know what young Graham Bell, toiling away in his couple of small rooms with his single assistant, would have thought of the group of immense laboratories that now bears his name, and which played so important a role in transatlantic telephony. At first sight, when one comes upon it in its surprisingly rural setting, the Bell Telephone Laboratories' main New Jersey site looks like a large and up-to-date factory, which in a sense it is. But it is a factory for ideas, and so its production lines are invisible.

The plural form 'laboratories' is correct, since the physical plant is in four separate locations, one in down-town New York and the other three in New Jersey. However, 'Bell Labs' is invariably used as a singular noun, like 'United States', so we will conform to this convention even though the resulting grammar may sometimes look a little odd.

Bell Labs is not unique, either in the United States or elsewhere, as nowadays many other great industrial organisations sponsor pure scientific research and what has been neatly called 'creative technology'. However, it is the largest entity of its kind, and probably the most famous. At the moment it has a staff of some seventeen thousand, of whom about seven thousand are scientists or engineers, and it costs its parent body, the American Telephone and Telegraph Company, a modest $600,000,000 a year.

AT & T can afford it. If most of us were asked to name the company with the largest capital assets in the world, we would probably plump for Ford or General Motors or Metropolitan Life. In fact, AT & T heads the list, its *annual revenue* alone adding up to the awe-inspiring total of twenty-three billion dollars ($23,000,000,000). Bell Labs is in small measure responsible for this.

The Laboratories has a divided allegiance, since half its stock is owned by the Western Electric Company – builders of most of the equipment for the enormous Bell Telephone System. Much of the Labs' work is concerned with design and development in the communications field – which today includes radio, TV, radar, missile guidance, and the whole explosively expanding empire of electronics. But its most important and most interesting activity is nothing more or less than discovery.

This is not something that can be planned and produced by a given delivery date. No executive vice-president can say: 'We'll have twenty basic scientific discoveries in the next financial year.' The only thing that can be done is to catch one's scientists (preferably young), pay them enough money to keep them from worrying about the rent, and give them pleasant offices where they can study whatever it is that interests them. This is expensive, and there is no guarantee at all that the results will be of the slightest commercial value, either today or a hundred years hence. But that twenty-three billion suggests that the gamble is well worth taking for any organisation that can afford it.

During the half a century since the Labs' formation in 1925 its workers have accumulated two Nobel Prizes and pioneered such revolutionary devices as crystal oscillators and filters, waveguides, and negative feedback amplifiers, each one of which has created entire new fields of electronics. Negative feedback, for example, was invented in 1930 and is now the principle upon which every hi-fi amplifier in the world is designed. And without waveguides, modern radar would be impossible.

In the realm of basic research, perhaps the most important of the hundreds of scientific discoveries flowing from the Labs was that of electron diffraction – for which Davisson won a Nobel Prize in 1937 – and cosmic radio noise, which Karl Jansky detected in 1932, and which would certainly have gained him a Nobel award if anyone at the time had had the slightest idea of its importance. The first discovery proved that the particles making up what is naïvely called solid matter have the property of waves; the second founded, a dozen years later, the vast new science of radio astronomy, which has revealed a new and unsuspected universe around us.

In recent years, the most dramatic example of the way in which research for its own sake can pay off to an extent beyond all computation was the discovery of the transistor at Bell Labs in 1948. This wonderful little device, which earned the Labs its second Nobel Prize, arose from fundamental research by Brattain,

Bardeen and Shockley into the way in which electricity flows through certain substances known as semi-conductors. These are materials (frequently crystalline) which, though much poorer conductors than metal, let electricity leak through them at a rate which takes them out of the insulating class. Sometimes they conduct better in one direction than in another; the classic example of this was the old crystal and cat's-whisker detector which was the heart of so many radios in the 1920s.

Though it made a rather unexpected come-back during the war in certain types of radar equipment, the crystal detector vanished completely from the radio field. It could detect signals, but it couldn't amplify them – and the electronic tube or valve could do both.

Then came the discovery that, in the right circumstances, certain types of crystal could amplify, and also possessed very great advantages over conventional tubes (extremely small size, low current consumption, absence of heating, ruggedness). The name 'transistor' was coined to describe such a device, and a revolution in electronics started which in a few years was to change the world. Its first impact was in the small but important field of deaf aids, which promptly shrank to invisible size and reduced their battery consumption to a fraction of its former value. Then came portable radios that were really portable, 'giant' computers that could fit into filing cabinets; and before long there will be more tiny transistorised devices watching over our safety, supervising our industrial processes, providing our communications and entertainment, than there will be human beings on the surface of this planet.

And it all started because three inquisitive scientists, for reasons best known to themselves, wanted to find out what happened when they passed an electric current through minute pieces of the obscure and unimportant element germanium.

Much of the excitement and stimulus one gets from a visit to the Bell Labs comes from the realisation that one is watching the birth of the future. It is impossible to guess which particular project will turn out to be of revolutionary importance, and which will amount to nothing more than an unnoticed letter tucked away at the back of the *Physical Review*. As examples of the sort of things the Bell Labs scientists get involved in, here are some choice specimens from over sixty papers listed in a single issue (January 1957) of the *Bell System Technical Journal*. Take a deep breath. . . .

'Quenched-In Recombination Centres in Silicon.'
'Observation of Nuclear Magnetic Resonances Via the Electron Spin Resonance Line.'
'Energy Spectra of Secondary Electrons from Mo and W for Low Primary Energies.'
'The Dipole Moment of NF_3.'
'Ballistocardiographic Instrumentation.'
'Ultrasonic Attenuation at Low Temperatures for Metals in the Normal and Superconducting States.'
'Refined Theory of Ion Pairing.'
'Effects of Superexchange on the Nuclear Magnetic Resonance of MnF_2.'
'A Developmental Intrinsic-Barrier Transistor.'
'Theory of Plasma Resonance.'
'Artificial Living Plants.'

To avoid fruitless speculation, perhaps I should explain that the last paper was a piece of very long-range thinking (by E. F. Moore in the *Scientific American* for October 1956), concerning the possible creation of mechanical 'plants' – for want of a better word – which would go foraging over land or sea, collecting and processing the materials needed for mankind, and reproducing themselves in the process. Moore is a colleague of Claude Shannon, another uninhibited thinker we shall meet in a minute.

Since this is an age, alas, of Secret Science, it is inevitable that a great deal of the work going on at the Bell Labs is highly classified. As you walk along the corridors past workshops, store-rooms, offices and laboratories, every so often you come across locked and sealed rooms bearing unwelcome notices, and sometimes patrolled by, surprisingly, unarmed guards. It is a fairly (though not completely) safe bet that what is going on in such places is not of fundamental scientific interest. Improving the intelligence of missiles, the security of communications systems or the accuracy of radar sets is of great military importance and may have valuable consequences in other fields, but in the long run what really matters is the work that seems to have no practical applications at all.

A Chinese writer once remarked that all human activity is a form of play. He would have considered this theorem proved beyond all doubt, at least as far as scientific activity is concerned, could he have accompanied me on my various visits to Bell Labs and watched its denizens entertaining themselves with some of the gadgets they had built.

Perhaps the most thought-provoking of these was Claude Shannon's mechanical mouse. It was at first sight somewhat surprising to find one of the most eminent mathematicians in the United States, and a founder of Information Theory (the mathematical basis of communication – using that word in its most general sense), playing with a small toy mouse that had obviously come from the local dime store, but there was a method in his madness.

Shannon's mouse is a highly sophisticated animal. It lives inside a metal maze, through the labyrinths of which it wanders in a random search-pattern until, by pure trial and error, it comes to the end of the maze and is 'aware' of its arrival when its whiskers close an electric circuit. If you then take it back to the beginning of the maze, it will now head straight to its goal in an apparently intelligent and purposeful manner, without making any mistakes or going up any blind alleys. To put its behaviour in anthropomorphic terms, the mouse has 'remembered' which of its blundering trials was successful, and has 'forgotten' all the others.

And what is the point of this? Well, to some extent the mouse represents the behaviour of an automatic telephone selector looking for the desired circuit once a number has been dialled. But the implications of the mouse are very much wider, for though its accomplishment is a relatively trivial one, it is the prototype of a *machine that can learn by experience*. It is thus something quite new, and not merely a robot that can do only what it has been told. True one might say that Shannon has 'taught' it to learn – but when he drops it into the maze it is on its own. And is the human brain anything more than a machine that can learn by experience as it blunders through the maze of life?

Machines that can mimic intelligent behaviour are not only of great philosophical interest and possible practical importance, but they are extremely stimulating (and frequently frustrating) to anyone who comes up against them. It is, for example, somewhat humiliating to be out-guessed by a pile of electronics the size of a small filing cabinet, as happened to me on my last visit to Bell Labs. This particular machine depends for its operation on the fact that a man is incapable of behaving in a completely random manner; everything we do has a pattern, conscious or unconscious. Thus if you are asked to call a random series of heads and tails, it is impossible for you to do so.

The machine I pitted myself against was told, by the pressing of the appropriate switch, whether I called heads or tails, and had to guess what my next call would be. When I tried to be clever and called a continuous series of heads, it took only three or four moves for my adversary to realise what I was doing and to predict that I would continue to call heads. When I switched back to tails it stayed in the heads groove for only a couple of calls before chasing after me.

In a short run, a man might beat the machine. In a sufficiently long one, however, he would have given it enough statistical information for it to predict his strategy. And there are implications in that last word ranging all the way from business through social relations to international politics.

After such esoteric devices, it might be a relief to mention two perfectly straightforward projects I encountered, which have immediate practical applications understandable by everyone. The first was a programme to see if anything can be done to improve the design of something that had been taken for granted by the whole world for a generation.

The familiar telephone handset seems about the ultimate in functional design. But nothing human is perfect, and there may be room for improvement here. If you had never seen a telephone before, but had merely been told what it had to do, how would you design it?

This was the question that a team at Bell Labs had asked itself, and it had produced dozens of answers. Some of them looked like anything but telephones; there were flower vases, pieces of abstract sculpture, salt-shakers, table cigarette-lighters . . . Perhaps the most interesting specimen had the calling dial built neatly into the handset itself, and not forming part of the base and cradle. Microphone, earpiece and dial formed one compact unit that fitted snugly in the hand; the Museum of Modern Art would have loved it. A decade from today, it may be one of the most familiar objects in the civilised world.

A second project had equally universal applications; it involved psychology, and an obscure branch of mathematics known as the Theory of Partitions. This may seem heavy artillery to bring to bear on a trivial problem, viz. how do you remember a telephone number?

Large cities like London and New York have so many sub-scribers that seven-figure numbers are necessary. In both cases, words are used as part of the identification, but they are purely an aid to memory, since each letter merely stands in for a number. The mechanism behind the dial knows nothing of letters—only the digits 1 to 0.

However, the number of reasonable names for exchanges is limited; sooner or later we may have to use nothing but digits, and perhaps even seven will be insufficient. When that happens, how is the non-mathematical man-in-the-phone-booth to remember such numbers as 3952841 or still worse, 96821473? A lot of people have difficulty in carrying the existing numbers from the telephone directory to the dial, and as for keeping them in their heads . . .

The answer seems to be that these long numbers must be broken up into segments; the problem is to decide where the break or breaks shall occur. A seven-digit number, it is rather surprising to find, can be split (partitioned) in no less than thirty ways – without, of course, altering the order of the digits, which would turn it into something else. To give an example, the number 1234567 can be written, spoken and – most important – carried in the head as

$$123\text{-}4567$$
$$12\text{-}34\text{-}567$$
$$123\text{-}4567$$

and in twenty-seven other ways you may care to work out for yourself.* The dash represents a verbal or mental pause, and only field studies and Gallup Polls can decide where the public prefers to have these pauses. A little reflection will show that this is very

* 1-2-3-4-5-6-7 doesn't count. After all, it's the same as 1234567.

far from a trivial matter; an incorrect decision could greatly increase the percentage of wrong numbers dialled and the general irritation among the telephone-using public.

I have just found among my own memories a perfect example of the way in which this partitioning works. Though I've not used my RAF airman's number for fifteen years, it is still available on demand because I stored it away in my brain not as 1097727 but as 109-77-27. Yet my shorter officer's number, though used over a longer and more recent period, has vanished completely; I remembered it as a complete six-digit entity, and now it would probably take hypnosis to bring it back.

So much, then, for a few of the hundreds of projects under way at any one moment at the Bell Labs. They are probably not representative, but merely happen to be ones I have encountered personally. However, they do give some idea of the immense range of activity and the general intellectual ferment which takes place when enough scientists are locked up together, with or without definite problems to tackle.

But before we go on to the specific project which is the main theme of this book, and which is perhaps the most daring technical feat yet attempted by the organisation, I cannot leave Bell Labs without mentioning one more device which I saw there, and which haunts me as it haunts everyone else who has ever seen it in action.

It is the Ultimate Machine – the End of the Line. Beyond it there is Nothing. It sat on Claude Shannon's desk driving people mad.

Nothing could look simpler. It is merely a small wooden casket the size and shape of a cigar-box, with a single switch on one face.

When you throw the switch, there is an angry, purposeful buzzing. The lid slowly rises, and from beneath it emerges a hand. The hand reaches down, turns the switch off, and retreats into the box. With the finality of a closing coffin, the lid snaps shut, the buzzing ceases, and peace reigns once more.

The psychological effect, if you do not know what to expect, is devastating. There is something unspeakably sinister about a machine that does nothing – absolutely nothing – except switch itself off.

Distinguished scientists and engineers have taken days to get over it. Some have retired to professions which still had a future, such as basket-weaving, bee-keeping, truffle-hunting or water-divining.

They did not stop to ask For Whom the Bell Labs Toll.

CHAPTER 20

Submarine Repeaters

When, in November 1953, the British Post Office, the Canadian Overseas Telecommunication Corporation and the American Telephone and Telegraph Company signed the contract to build the first transatlantic telephone cable, both parties already had many years of experience with submerged repeaters of various types, though in circuits far shorter than the one now proposed. The Post Office had laid a repeater in the Irish Sea, between Anglesey and the Isle of Man, as early as 1943, after five years of experimental work. This was followed by other repeaters in telephone cables to the Continent, but all of these were laid in fairly shallow water, and could not have withstood the enormous pressures existing at the bottom of the Atlantic.

In the United States, on the other hand, interest had been focused from the beginning on repeaters which could be laid in the deep ocean. Back in the 1930s, as we have already mentioned in Chapter 16, advances in electronics had caused the Bell System to think seriously about transatlantic cables with submerged repeaters, and a great deal of experimental work was done before and during the second world war on the development of the very reliable components which would be needed.

This work culminated in 1950 with the laying of a telephone cable between Key West (Florida) and Havana (Cuba), a distance of about 140 miles. Six repeaters were used, some of them at depths of over a mile. From the beginning, this cable was regarded as a model of the proposed Atlantic cable, and its performance was therefore watched with extreme care. When it had given two years of trouble-free service plans went ahead on the far more ambitious project, and consultations were started between the technicians of the British Post Office and AT & T.

At this point, a fundamental difference in approach became obvious which determined the entire design of the transatlantic telephone cable, and also provides an encouraging example of the way in which international conferences can succeed if only the politicians can be kept out of them. To understand why this difference arose, it is necessary to state what any submerged repeater has to do, and how it does it.

A repeater is simply an amplifying device consisting of several vacuum tubes and their associated circuits, which boosts a faint incoming signal (by a factor which may be as much as a million) and sends it on its way into the next section of coaxial cable. It has to be sealed in a completely watertight container, and a casing which can stand the pressure at a depth of two and a half miles is both massive and heavy. Yet it has to be paid out with the cable as the ship steams along like a spider spinning out its thread.

The British shallow-water repeaters are encased in stubby, rigid tubes about the size and shape of torpedoes, and the cable ship has to be brought to a full stop when it is time to lay one. This does not matter in shallow water, but it has long been known that when a great length of cable is being paid out, as would be the case in the open Atlantic, stopping the ship introduces very serious danger of kinking. The spiral wires which provide armouring tend to untwist slightly when the cable – perhaps fifteen thousand feet of it, weighing many tons – hangs freely down from the ship. When the paying-out is steady and continuous, this untwisting spreads itself uniformly along the cable without causing any harm; but stopping the ship can result in kinks which may distort the cable or even, in extreme cases, tie it in knots. It is quite astonishing what a piece of cable which looks as rigid as an iron bar can do during its brief period of freedom on the way from the ship's tanks to the sea-bed; sometimes knots are produced that could hardly be bettered by a kitten playing with a ball of wool.

To prevent such disasters, the Bell System engineers designed repeaters which were flexible, so that they could not only be paid out with the cable but would be virtually part of it. A section of cable holding a repeater looks rather like a boa-constrictor after a light lunch; only a barely perceptible bulge shows that anything out of the way has taken place.

It is hard enough to build electronic equipment that can function on the sea-bed for decades without attention; packing it into a tube an inch across, and capable of bending round a drum only

L

seven feet in diameter (the size of the paying-out sleeves on the cable ship) makes matters even more difficult. We will see later how it is done; the important point at the moment is that the extremely small diameter of the deep-water flexible repeaters means that they can transmit signals *in one direction only*. There is simply not room for the equipment that would allow two-way working.

The much bulkier British repeaters, on the other hand, can work in both directions. They also contain two identical parallel amplifiers, so that if a component fails in one, the unit as a whole will still continue to operate with practically no loss of efficiency. The Post Office design is much more modern than that of the Bell Labs, which has been deliberately based on well-tried vacuum tubes that have given good service for many years. Nothing has gone into the flexible repeaters that has not proved itself as nearly completely reliable as any tests can show. The philosophy of the Bell engineers has been that of a man with a very important journey to make, who has decided that the best way of arriving safely is to build a careful copy of a 1935 Rolls-Royce and to drive it at twenty miles an hour. This is not excessive caution, for the failure of a single tube, resistance or even soldered joint in one of the deep-sea repeaters, might easily cost a hundred thousand pounds in repair-ship time and lost revenue.

Faced with these two different approaches, a neat compromise was reached. The Bell repeaters would be used for the Atlantic crossing, even though this would mean laying two cables – one to carry speech from east to west, the other to carry it from west to east. The British Post Office repeaters would be used in the shorter and shallower sea-crossing from Newfoundland (where the trans-ocean cables come ashore) to the Nova Scotia mainland, and because they could amplify signals passing in both directions only a single cable need be laid in this section. From Nova Scotia the service would be continued to Canada and the United States by land line and microwave links.

Once this basic decision had been made, the second problem (which was of course bound up with it) was the route to be selected for the cable. At that time, there were no less than twenty tele-graph cables across the Atlantic, and it was essential to give existing circuits as wide a berth as possible. Most of these head out across the Atlantic from the south-west tip of Ireland, and it was decided to use a route which would be far north of them for most of the way. The cable would start from the little town of Oban, in Scotland, and would thus by-pass Ireland completely.

At the eastern end, an attempt was made to avoid the cat's-cradle of existing telegraph circuits off Newfoundland by bringing the ends ashore at Clarenville, a few miles north of the fourteen other cables which converge on the island. The cable through which America would talk to Europe would be laid first; its twin, which would carry speech from Europe to America, would be about twenty miles to the north, so that if necessary one could be grappled and hauled up without disturbing the other. The greatest depth encountered on the route would be about thirteen thousand feet.

The remaining technical details were worked out in a series of conferences between the parties involved – the British Post Office, the American Telephone and Telegraph Company, and the Canadian Overseas Telecommunication Corporation. Despite all the quite unprecedented problems involved – some of which will be touched upon in Chapters 21 and 22 – the entire £14,000,000 project was completed in three years, beating the target date by several months.

It would be very hard to find another enterprise of a similar scale which was carried out on such an international – one might say supranational – basis. The cost was shared jointly between Britain, the United States and Canada, mostly in terms of services or equipment supplied. Perhaps the most expensive single items were the 102 deep-sea repeaters in the Atlantic, built in a special factory set up by Western Electric at Hillside, New Jersey. With the exception of 133 miles manufactured by the Simplex Wire and Cable Company at Portsmouth, New Hampshire, and the sixty miles made by W. T. Henley's Telegraph Works Co. Ltd. for the section across Newfoundland, all the cable for the project was produced at Erith, Kent, by Submarine Cables Ltd., a direct descendant of the firm that made the first Atlantic cable a hundred years ago.

The British Post Office had two main responsibilities. It provided the shallow-water repeaters for the Newfoundland–Nova Scotia link, and the cable ship to do the laying. Her Majesty's

Telegraph Ship *Monarch* is the largest cable ship in the world, being a worthy (and more economical) successor to the fabulous *Great Eastern*. No other vessel could have done the job so efficiently, and even so the *Monarch* was forced to lay each of the deep-sea sections of cable in three instalments.

The total list of credits in the project is at least as impressive as in any Hollywood super-epic, but at the risk of hurting a good many people's feelings we will go on to consider what was done, not who did it. As in most great engineering undertakings today, there is no single individual to whom one can say: 'This is the man who made it possible.' Much of the responsibility for making the original technical decisions must go to the president of the Bell Telephone Laboratories, Dr. Mervin J. Kelly, and to the Director-General, and former Engineer-in-Chief, of the British Post Office, Sir Gordon Radley, KCB. They were the two men who would have had to face the music if the project had failed, but the actual designers and builders of the system were scores – no, hundreds – of engineers and scientists of whom the public will never hear. Their satisfaction lies in the knowledge of a worth-while job well done, and perhaps a paper explaining how they did it, read to their colleagues and then published in the technical literature.*

We can consider only the highlights of the project, ignoring all sorts of fascinating problems and their ingenious solutions, which would take an entire book to deal with and even then would be of interest only to specialists. (Though as that famous craftsman Chick Sale so amply proved, there is no specialist who cannot make his subject interesting if he knows how.) We will look first at the heart of the entire system – the flexible repeater which makes the transatlantic link possible.

There are 102 of these repeaters now lying far down on the bed of the Atlantic, fifty-one in the eastbound and fifty-one in the westbound cable. They are spaced about forty miles apart, and they are some of the most remarkable pieces of precision machinery ever made.

* The technical details of the transatlantic telephone cable were published simultaneously in a series of papers in the *Bell System Technical Journal* and the *Journal of the Institution of Electrical Engineers* for January 1957, from which much of the material in the following chapters is taken.

Electrically, they are very simple – scarcely more complex than the average radio set. Each contains three vacuum tubes of a specially developed tough and rugged type, and about sixty resistors, capacitors and other circuit components. In the ordinary way, there would be no particular difficulty in building such a unit so that it would give several years of reliable service.

That, however, would be nothing like good enough. Each repeater is a link in a chain; if a single one fails, the whole chain is useless. In the east–west and west–east pair of cables, there are altogether 306 vacuum tubes and some six thousand other components; to make reasonably sure that the system will operate for the planned twenty years without failure, the degree of reliability required for each item is fantastically high. Luckily most electrical parts do not wear out; the average resistor or inductor, if well made and not overloaded, will last virtually for ever.

Each amplifier has to handle a band of frequencies broad enough to contain the thirty-six simultaneous telephone conversations passing through the cable. The band used is, by radio standards, a very low-frequency (long wave) one; no domestic receiver could tune down to it. The range actually employed is from 12 kilocycles to 174 kilocycles per second; the human ear could easily detect the lower end of the band as a high-pitched whistle.

The thirty-six channels are also packed much closer together than are the stations on your radio dial. This can be permitted because the standards required for telephoned speech are a good deal less rigorous than for music. But, if necessary, two or three of the telephone channels can be combined to provide a single high-quality music channel.

To make up for the losses in the forty miles of cable leading in to it, each repeater has to amplify the incoming signals approximately a million times, and this brings us to what is undoubtedly the most awe-inspiring statistic in the whole enterprise. Since there are fifty-one repeaters in the circuit, this means that the *total* amplification along the line is given by the colossal figure of *a million multiplied by itself fifty-one times* – or 1 followed by 306 zeros!

Let us pause to contemplate this number for a moment. It would be a gross understatement to call it astronomical – there is no quantity, anywhere in the natural cosmos, that begins to compare with it in magnitude. The number of grains of sand on all the shores of earth? That's too small even to bother about. If the whole world were made of sand, the total number of grains could be written out in half a line of type; there would only be about thirty zeros in it, not three hundred. The number of electrons in the entire cosmos? Well, that's a little bigger; there may be as many as eighty digits in it – but that's still not within hailing distance of ten to the power of 306.

This truly stupendous number appears twice in the mathematics of the Atlantic telephone cable. It is not only the total amplification (or gain) produced by the repeaters, but also the total loss (or attenuation) along the line, which these repeaters have to counteract. So the engineers have had to perform a kind of balancing act, designing and adjusting the overall circuit so that the losses precisely equal the gains. You will appreciate now why there was no possibility of establishing a multi-channel transatlantic telephone cable without repeaters. The total energy output of all the stars in the cosmos would not be sufficient to give a measurable signal, after being divided by ten, 306 times in succession.

Since electron tubes require current to operate their heaters and anode circuits, a d.c. supply is provided by applying two thousand volts (of opposite polarities, so the available voltage is four thousand) to either end of the cable's central conductor. Very elaborate precautions have been taken to guard against the risk of power failure, and also against naturally produced surges of electricity ('earth currents') which might damage the repeater circuits.

We will have more to say about the electronics of the system, for those who wish to go into it in a little more detail, in Chapter 21. Equally important, and perhaps of more general interest, are the purely mechanical aspects of the design. Just how does one pack three vacuum tubes and sixty other assorted components in a narrow pipe which will bend round a seven-foot-diameter circle, yet will withstand a water pressure of three tons to the square inch?

The Bell Labs answer looks rather like a piece of bamboo. Each repeater consists of seventeen short tubes of transparent plastic, averaging six inches in length, flexibly coupled together at the ends so that the whole assembly can bend like a string of rail-road coaches going round a curve.

To provide the necessary strength, the series of plastic cylinders is enclosed inside a double layer of steel rings, which overlap each other and give a tough yet flexible armouring which would have created quite a sensation at the Court of King Arthur. On top of this is a long copper tube – and over that are the multiple layers of jute bedding and steel wires which give protection for the rest of the cable. As each repeater was completed and passed its tests, it had to be taken to the cable works where the normal armouring was spun over it. A repeater ready to be dropped on to the sea-bed looks like an unusually stout piece of wire cable, some eight feet long, which slowly tapers off from two inches down to one inch in diameter. There is no indication at all that the central bulge contains electronic equipment costing some £20,000. Why the bill was so high will be obvious when we see the gauntlet of tests that every component had to pass – tests so rigorous that sometimes less than 1 per cent of the specimens got through.

Quite extraordinary precautions were taken in the packing and shipping of the repeaters once they had left the factory. They were nestled like eggs in thirty-four-foot-long containers, and travelled to the airport in special trailers which could be heated or cooled so that the specified temperature limits would not be exceeded. Then they were flown to England to be spliced into the main cable and loaded aboard the *Monarch*.

The cable itself, though a fairly straightforward piece of engineering compared with the repeaters, was manufactured according to the most rigorous specifications ever laid down. It might be expected that the strongest and most heavily armoured section of a submarine cable would be that which is laid in the deepest water, but this is not the case. Once a cable has been safely deposited on the sea-bed, it is beyond the reach of almost any agency that might damage it. The cables that need protection are the shallow-water ones, and those that actually come ashore are the most heavily armoured of all, since they have to withstand the assaults of marine animals, waves and trawlers.

The deep-sea section of the Atlantic telephone cable is about one and a quarter inches in diameter, and consists of ten separate layers from the central copper wire to the outer coating of protective jute yarn. It is interesting to note that the cable contains one primitive and possibly useless survival (like the human appendix) from earlier stages of evolution. Many of the first telegraph cables – particularly those in shallow and tropical waters – were put out of action by the voracious appetites of marine borers which developed an unexpected taste for electrical insulation. To frustrate these minute menaces, thin copper tapes ('teredo tapes') were wound over the insulation.

It is most unlikely that any such borers exist at the bottom of the Atlantic, but the telephone cable still has a teredo tape, retained on the principle that it was safer to leave it in than take it out.

Apart from the obvious importance of using good-quality materials, so that the electrical losses would be low, it was essential that the cable be as uniform as possible. Any irregularities – caused, for example, by the central core being slightly eccentric, or by a variation in diameter of either conductor – would produce reflections in the line, so that part of the transmitted energy would bounce back the way it came.

The production of this cable within less than two years was a truly remarkable feat of engineering. The deadline was set by the fact that owing to uncertain weather no cable-laying can be attempted in the North Atlantic after late August; a delay of a week or two in delivery might, therefore, cause the whole enterprise to be postponed for an entire year. Submarine Cables Ltd. had to build a brand-new factory beside the Thames, dredge a berth in the shallow river for the *Monarch* so that she could come alongside for loading, and then start production of more than four thousand miles of cable to standards of perfection never before attempted. Work had to continue day and night, through week-ends and public holidays in the race against the clock.

Had it not been overshadowed by the more glamorous and technically more daring transatlantic circuit, the link from Newfoundland to the Nova Scotia mainland would itself have been one of the world's most outstanding feats of communication. This Clarenville to Sydney Mines cable employs sixteen of the British type of submerged repeaters, built by Standard Telephones and Cables Ltd. Their amplifiers deal with the thirty-six west–east telephone circuits within one band of frequencies (20–260 kc/sec) and the east–west ones in another (312–552 kc/sec), so that there is no possibility of interference. After amplification, the two groups of conversations are sorted out by electrical filters and routed east or west as the case may be. If this sounds a little complicated, it is very similar to what happens to traffic at clover-leaf intersections. Indeed, a multi-lane super-highway provides quite a good analogy to the Atlantic telephone system. In the Newfoundland–Nova Scotia link the thirty-six lanes in each direction are side by side in the same cable, whereas in the open Atlantic they are twenty miles apart in separate conductors. The filters in the British repeaters act like a clover-leaf arrangement, switching one flow of traffic so that all seventy-two streams pass through the amplifier in the same direction.

For part of its length, this section of the cable is overland. About sixty miles pursue a somewhat circuitous route across Newfoundland, from Clarenville on the eastern side to Terrenceville on the west, where the cable goes to sea once more on its last lap to America. As the land section had to be laid through bogs and ponds, often in freezing weather, the installation teams must have envied the crew of the *Monarch*, who merely had to pay out the cable over the sheaves as the ship ploughed effortlessly across the Atlantic.

Perhaps at this point, before most readers are completely blinded by science, it may be a good idea to sum up this chapter by considering the adventures and transformations which the human voice experiences on its way from, say, New York to London.

First, there is the pressure wave in the air caused by the speaker's larynx and vocal cords. This hits a microphone diaphragm, behind which are packed some grains of carbon through which a weak electric current is flowing. The varying pressure on these granules by changing their resistance, causes fluctuations in this current. The speech waves have been turned into electric waves of the same frequency (between approximately 300 and 3,000 c/s) which pass through the local Bell System telephone exchange and are routed to a switching centre at White Plains, thirty miles north of New York.

At this point the waves enter their first coaxial cable, sharing it with some dozens of other conversations. They have now been transformed into something above the range of hearing, being carried on a high-frequency wave in exactly the same way as speech or music rides a radio wave through the ether.

Three hundred miles later, at Portland, Maine, they are in fact launched into the ether along a microwave link, jumping from tower to tower via tightly beamed radio waves at the very high frequency of four thousand mc/s. (That is, a wave-length of about eight centimetres, which is similar to that used by many radar installations.)

The microwave chain crosses Maine, New Brunswick and Nova Scotia for almost six hundred miles until it reaches the sea at Sydney Mines, facing Newfoundland across the Gulf of St. Lawrence. Now the words go underwater, and at the same time the waves that carry them are stepped down to a more reasonable frequency – somewhere between 20 and 260 kc/s.

For 376 miles they pass along a single coaxial cable, being amplified by sixteen of the British two-way repeaters located at approximately equal intervals along the line. They take no notice at all of the thirty-five other conversations going in the same direction – or the thirty-six passing them in the other, but spaced much higher up the spectrum of frequencies.

At the lowest point of their travels the words move through waters 1,200 feet deep, and for the last sixty miles of their journey they go across land that is so wet that they might just as well remain underwater. Indeed, the repeaters in the Newfoundland link are lying at the bottoms of ponds.

Now they are ready for the big jump. At Clarenville, still riding on a carrier wave in the low-frequency-radio range, they enter the southern twin of the transatlantic cable. For 2,240 miles, descending to depths of almost fourteen thousand feet, they travel along the bed of the Atlantic. Fifty-one times they fade away, much as the spoken word fades with distance – and fifty-one times they are amplified a millionfold and sent refreshed on the next stage of their journey.

Five hundred miles from the British Isles they climb the sunken mountain range of the Rockall Bank and come to within three thousand feet of the surface, but soon sink again back into deep water. Two hundred miles off Scotland they hit the edge of the continental shelf, rising first swiftly, then slowly up from the depths to enter the Firth of Lorne and come ashore at the quiet little town of Oban.

There are still 550 miles to go, but the main adventure is ended. It's back to the humdrum coaxial through Glasgow and across the border down to London and the end of the line. The carrier wave is suppressed, and only the audio-frequency oscillations are left, to be turned back into speech by a magnet and vibrating diaphragm arrangement almost unchanged since Graham Bell invented it a century ago. . . .

This blow-by-blow commentary has taken two or three minutes to read, and what has been left out would fill a book many times the size of this one. Yet the whole process takes much less than a tenth of a second; we have indeed gone a very long way from the time when it required sixteen hours to tap out Queen Victoria's message of greetings to President Buchanan.

CHAPTER 21

Technical Interlude

This chapter is intended primarily for readers who have some knowledge of electronics and would like to have a few more details about the problems met and solved in the design of the transatlantic telephone system. At the same time, much of the material that follows should not be completely beyond the reader who has no technical background, if he cares to plough hopefully ahead.

As far as the submerged repeaters are concerned, the most difficult problem is of course reliability. Sooner or later a repeater will fail and will have to be replaced, but the cost of cable-ship time and lost operating revenue will be at least a hundred thousand pounds. The target aimed at is a service life of twenty years for the overall system, and in the quest for this goal an elaborate programme of tests has been carried out on every item which might conceivably cause trouble. Even so, one possible source of disaster was discovered, almost accidentally, by a telephone maintenance man who had nothing at all to do with the project but was intelligent enough to report what he had found.

This was the discovery of 'whiskers' – something which no one would ever have dreamed of looking for. Under certain conditions tin (and other metals) will grow fine, hairlike threads which may easily bridge a narrow gap and cause a short-circuit. It is quite remarkable that this effect has been discovered so recently (1951) when one considers how long the world has been using tin. Probably hundreds of men had noticed it and, thinking the growth was some kind of fungus, had rubbed it off and ignored the matter. It was fortunate that the discovery was made before the completion of the telephone cable; as it was, entire batches of special vacuum tubes had to be scrapped because they had been made with tin-plated leads.

This, incidentally, is not the only example of the odd behaviour of tin. At low temperatures it is liable to turn from solid metal into crumbling powder, and the deaths of Captain Scott and his companions on their return from the South Pole have been attributed, at least partly, to this fact. The cans of kerosene, which they had cached along their return route, were found to be empty when they uncovered them. The tin solder had crystallised in the protracted cold and the precious fuel had escaped.

Some of the elaborate precautions taken against mechanical and electrical failures will be mentioned in the next chapter; they involved rigorous selection of materials and surgical cleanliness in the manufacturing and assembling operations. But no amount of forethought in design, and care in its execution, can guard against weaknesses which only time will reveal. It was fortunate, therefore, that during the 1930s the Bell Labs had begun a series of life tests on vacuum tubes to discover, and if possible eliminate, their causes of failure. These tests were part of a long-term policy like that of laying down a cellar of wine for one's children; when the transatlantic telephone was being planned, some of the vacuum tubes under test had been running continuously for seventeen years. They gave the designers confidence that the enterprise had a good chance of succeeding.

This slow piling-up of operating hours, during which batches of tubes were run under various loads and at different voltages and cathode currents, explains why the tubes used in the Bell repeaters are of a mid-1930 design. And also why, as explained in more detail in Chapter 23, the new transistors could not be used.

The three tubes used in each of the 102 flexible repeaters in the transatlantic circuit are pentodes (175 HQs) operating with a plate voltage of 50, and a heater voltage of 18. These voltages are obtained from the total drop along the cable's central conductor, which has a four-thousand-volt potential difference between the Newfoundland and Scotland ends.

Since the 306 tubes forming the complete amplifying chain are in series, if a single one fails the whole system will cease to operate. Sooner or later one *will* fail; the problem then will be to discover where it is along two thousand miles of Atlantic sea-bed. With a simple, unrepeatered cable, methods of locating breaks have been known for many years, but these are not applicable in a case such as this. Built-in testing circuits have been incorporated into the repeaters, so that each one signals to the shore stations that it is functioning properly.

The method used to do this is highly ingenious. Every repeater contains a crystal, very sharply tuned to a frequency above the band used for transmitting signals. These crystals are in a feedback loop, so arranged that at their frequency the amplifier gain 'peaks' violently, by some 25 db. This results in each amplifier producing a narrow band of noise, which can be detected by measuring equipment at the shore ends.

The narrow noise peaks (about 100 c/s wide) can be tuned in one by one, and each of the 102 corresponds to a known repeater. If all the peaks are present and correct, then every repeater is functioning correctly. However, if a repeater fails, then the receiving station will pick up only the noise peaks from the units between it and the point where the fault has occurred. The repeaters beyond this point may still be working, but their signals will not get through the dead section.

It will be noticed that there is one large assumption in this argument. For the test to work, the line as a whole must still possess electrical continuity, so that all the remaining repeaters can still function. But as all the 306 tubes are in series, if one fails – through the opening of a heater circuit – *all* the rest will go dead through power interruption.

To guard against this, each repeater includes a gas-discharge tube which will fire and by-pass the unit in the event of failure. Current will thus continue to flow and the remaining fifty repeaters will still be live. The system as a whole will, of course, be unable to transmit speech, but the testing circuits will still function so that the position of the fault will be accurately known.

One interesting consequence of the plans made to test the individual repeaters is that they may give us our first precise knowledge of the temperature across the bed of the Atlantic. Each of the very accurately tuned crystals in the test circuits is slightly affected by temperature changes, and though the change is only half a cycle (out of some 170,000 cycles!) per second for each degree Fahrenheit, this can be detected. As a quite unintended by-product of the transatlantic telephone we now have a string of thermometers between Europe and America, and their readings will be of extreme value to geophysicists.

There are so many hi-fi enthusiasts around nowadays that some details of the repeater circuits may be of interest. Each unit consists of a three-stage amplifier using the 175HQ pentodes already mentioned, connected to the cable through input and output coupling networks. There are two feedback loops which – together with the coupling and interstage networks – control the frequency-gain characteristics of the amplifier.

Unlike a hi-fi set, the amplifiers do not have 'flat' response curves. They are designed to give much greater gain at the high frequencies, to make up for their far higher losses in the cable. The gain produced varies from about 20 db at 20 kc/s to 60 db at 160 kc/s.

Despite all the calculations and tests that can be made on land, by the time a repeater and its associated cable has reached the sea-bed the transmission characteristics will have altered slightly from the desired value. Some of these changes are due to temperature or pressure, but the complete explanation of this so-called 'laying effect' is not yet clear. Because even small alterations in perfor-mance can produce drastic final effects after so many stages of amplification, they have to be neutralised during the cable-laying process.

The normal way of dealing with these variations would be to adjust the next amplifier in the chain, altering its gain versus frequency characteristics as desired. But all the amplifiers are completely sealed and buried inside the cable armouring; there is no way of getting at them once they have left the factory.

To deal with this problem, another type of submerged circuit element was developed – the equaliser. This is merely a simple network of resistors, capacitors and inductors, sealed up in a housing similar to that of the repeaters, but shorter. As the cable was laid, equalisers were spliced in at appropriate spots as found by measurement; altogether there are fourteen of them in the two Atlantic cables. Since *Monarch* could not be stopped during the laying process, and it took up to nine hours to make the neces-sary splices, inserting the equalisers was not as simple a matter as it may sound. Just what it involved will be seen when we describe the cable-laying operations in Chapter 23.

The British repeaters, in their massive pressure housings nine feet long and ten inches in diameter, were, as might be expected, a good deal more elaborate than their one-and-three-quarter-inch diameter American counterparts. The three electron tubes

employed in each amplifier are 6P12 pentodes, developed at the
Post Office Research Station at Dollis Hill. They have six times
the trans-conductance of the Bell Labs' 175HQs, but this has been
obtained at the cost of closer electrode spacing and therefore
greater possibility of failure. As insurance against this, each
repeater contains two identical amplifier chains in parallel, so that
if one fails the unit will continue to function. Many of the components in each amplifier are also doubled up for the same
reason; for example, the interstage couplings have two blocking
capacitors in series, so that if one breaks down there will be no
danger of anode voltages reaching the grid of the next tube.

As in the Newfoundland–Scotland section, the repeaters in
the Newfoundland–Nova Scotia cable contain ingenious fault-
locating devices. One of these depends upon what is virtually an
application of radar; a pulse is sent along the line, and each repeater
contributes a distinctive echo. This technique would be impossible
on the transatlantic section, where the cables transmit only in one
direction and so no echo could be returned.

This is perhaps as far as one can go, in a book not intended for
specialists, in describing the circuit details of the system. Those
who want to go into the matter further should have no difficulty
in getting hold of the relevant papers (see footnote, page 164).

There still remain, however, two important aspects of the enterprise which have been barely touched, and which merit chapters
on their own. The first concerns the extraordinary production and
manufacturing methods that were used in the relentless quest for
reliability; and the second takes us out to sea, aboard HMTS
Monarch as she shuttles back and forth between Scotland and the
United States, throwing her priceless cargo overboard.

CHAPTER 22

Production Line

Personally, I have never been much interested in the way in which things are made, as compared with the way in which they work. The processes whereby lumps or sheets of metal are converted into kettles, automobiles, machine-guns, ashtrays, doorknobs, knuckle-dusters and the other appurtenances of our civilisation have always been something of a mystery to me. I suspect that a surfeit of wartime documentary films showing the ingenious methods by which salvaged saucepans were turned into Spitfires is responsible for this mental blank.

In the case of the transatlantic telephone system, however, some of the manufacturing processes and precautions were so unusual that no account of the project can possibly leave them out. The search for reliability set new standards which may well have consequences in quite different fields.

There are very few, if any, products in the modern world which are delivered to the customer with a guarantee of perfection, though some commodities are more perfect than others – the spectrum of reliability ranging from parachutes at one end to certain types of patent can-opener at the other.

Most reputable firms have standards of inspection and control sufficiently high to ensure that the number of faulty items rolling off the production line is so low that it has negligible bad effect on customer goodwill. A firm that tested every single unit it put out, for every flaw that it could possibly develop, would not stay in business for long. It would have more employees testing its products than making them.

But when one is aiming to build a complex piece of equipment which is not accessible for servicing – save at enormous expense – ordinary manufacturing economics go by the board. If the job can be done only by having a dozen inspectors to every workman, then the bill has to be met. In the long run, the price may be small. A resistor which costs a shilling in the normal way might

M

cost ten times as much after it had been through its tests, and its below-standard companions had been eliminated. But as the failure of that single resistor could easily result in a bill for a hundred thousand pounds – the cost of raising and replacing a faulty repeater – it will be seen where the true economy lies.

The American flexible repeaters were built in a special plant constructed by the Western Electric Company at Hillside, New Jersey, to standards of cleanliness probably never met before outside the pharmaceutical or atomic energy fields. The whole factory was, of course, air-conditioned and temperature-controlled, the pressure inside being greater than that outside to provide an additional barrier to dust.

The employees were filtered almost as carefully as the air. They had to pass rigorous mental and physical tests, such as 'capability of performing tedious, frustrating and exasperating operations against ultra-high-quality standards'. Then they were thoroughly trained for their various jobs and, what was probably equally important in the long run, fully briefed concerning the complete project and its progress, by means of occasional talks and film shows. Even those who had nothing to do with the actual manufacturing operations – the clerical workers, watchmen, chauffeurs, boilermen and so on – were kept in the picture to give them the necessary team spirit. As a result, labour turnover was very low and attendance exceptionally good.

Because dirt carried on clothing was a possible source of contamination, all employees were provided with special Orlon uniforms and caps, which they had to put on when entering the working area. Anyone seeing the men and women of the Hillside plant on their way to work would have been quite certain that they were surgeons and nurses on the way to the operating theatre.

Almost all the components that went into the deep-sea repeaters were manufactured in this one plant, so that their entire life history was known. The only parts contracted out were items whose manufacture would have required the installation of very expensive machinery; the protective steel rings and copper tubes were examples of these.

One might imagine that something as straightforward as a ten-foot-long copper tube one and three-quarter inches in diameter and with a $\frac{1}{32}$-inch wall was fairly easy to obtain from a subcontractor. However, the tolerances required were so severe that only one supplier could be found – and he would only guarantee the tubes on the outside, not the inside as well. After he had over-hauled his equipment and done his best, *less than 1 per cent* of the tubes supplied met the specifications. Fortunately, further consultations reduced the rate of rejection from a ruinous 99 per cent to a tolerable 50 per cent.

This was an exceptional case, but even after the greatest care had been taken in selecting materials the rate of wastage was very high on most of the components going into the repeaters. Only 80 per cent of the inductors, 65 per cent of the resistors, and still fewer of the capacitors constructed in the factory with the utmost care finally made the grade.

Many of the manufacturing and assembling operations had to be carried out under magnifying glasses or even binocular micro-scopes; some tiny coils took a day and a half to wind by hand, no machine in existence being capable of doing the job.

The paper-work involved in the manufacture of the repeaters was as impressive as anything else on the project. A complete genealogy of every item, through all its inspections, was kept, and by the time the factory had turned out a hundred repeaters it had also produced about a thousand volumes of records and data. Though this may seem somewhat lavish, this small library will be invaluable during the twenty or more years during which it is hoped that the system will operate. When things start to go wrong, as sooner or later they must, these voluminous records will reveal why, and will prevent the same mistake being made next time.

Once the electrical components were wired up and embedded in the seventeen transparent plastic tubes which, linked together, make up a repeater circuit, they had to be sealed in their protective armouring. It is not a simple job to design and construct a flexible enclosure which will not admit one drop of water after twenty years under a pressure of three tons to the square inch, and some most ingenious methods were used to test the various glands and seals incorporated in each repeater.

One of the most delicate of these tests involved applying helium gas, at a pressure 25 per cent higher than that found on the bed of the Atlantic, to one side of the seal, and attaching an extremely sensitive detector known as a mass-spectrometer to the other.

Helium atoms, because they are very much smaller than water molecules, can pass through cracks or pores far more rapidly, and the detector employed could observe a rate of leakage corresponding to about one thimbleful of gas in fifty years. Even if water crept through the seal at a corresponding rate, a desiccator inside the cavity of the repeater would easily absorb it.

This desiccator unit, incidentally, had to remain unopened until the entire repeater had been dried – and sealed. But how could one make sure that it had opened, when there was no longer any way of observing it? The neat and simple answer was to strap a sensitive microphone to the wall of the repeater. When the desiccator can was forced open by gas pressure, the sound of the bursting diaphragm could be picked up as a faint, clear 'ping' through the steel armouring.

Another test for leaks involved an apparently impossible feat. After the last seal had been made and there was no longer any access to the inside of the repeater it was necessary to check that the final brazing operation was sound. This meant that, somehow, the last seal had to be tested *from one side only*, since there was no longer access to the other.

Radioactive tracers provided the answer here. A solution of a salt of the radioisotope Caesium 134 was applied to the seal for sixty hours at a pressure of several tons to the square inch. If any of the isotope leaked into the interior, Geiger counters could detect it through the repeater walls by the gamma rays it emitted. The Caesium 134 atoms would act rather like an army of secret agents, each carrying a radio transmitter, working its way across a heavily guarded frontier. Headquarters would be able to tell just how many – if any at all – had been able to get through.

This radioactivity test involved so many operations (including sixty precisely timed washings) that a recording of it was made on magnetic tape. Whenever the test was made, the recorder was switched on and the instructions followed as they emerged from the speaker. It was an excellent way of making sure that the timing of the operations was correct, but the testing crew must have been a little tired of that recording when they had heard it for the hundredth time. One hopes that suitable background music was provided at the more routine sections.

The British Post Office rigid repeaters used on the Newfoundland–Nova Scotia link of the system were made under very similar conditions at the Standard Telephones and Cables factory at Woolwich, London. Air-conditioning and extreme cleanliness were again the order of the day; to give some idea of the precautions taken, all test data were recorded on special paper which was not liable to fray into fluff or dust. The smock-clad workers christened their spotless plant the 'dairy', though one wonders how much milk has ever been processed in such fanatically hygienic surroundings.

Many of the parts of the British repeaters were gold-plated to avoid the 'whiskers' which tin is liable to grow, and it is interesting to note that two of the most valuable elements known to man – gold and radium – were incorporated in the telephone system. The radium was used (a millionth of a gramme at a time) in the American repeaters to promote quick operation of the gas-tubes which might have to shunt or by-pass a unit in the event of its failure.

These examples are, perhaps, sufficient to give some idea of the fantastic care taken, at every stage from selection of raw materials to final assembly of the finished article, in the production of the 102 American and sixteen British repeaters in the complete transatlantic system. It is safe to say that, apart from external catastrophes, not one of the units will ever have to face such severe operating conditions as it met during its tests. The stable, unchanging environment of the ocean bed is a nice, peaceful place compared with the mechanical torture-chambers in which the repeaters spent the last months of their existence on land, being tested at pressures up to five tons to the square inch.

Yet despite all this care and endless inspection (even the inspectors had a corps of super-inspectors who inspected them) nothing man-made can be wholly perfect, especially when thousands of different and often exceedingly delicate components are involved. The Atlantic telephone system must contain some unknown weakness, some heel of Achilles which will ultimately put it out of action. At this very moment a vacuum tube is slowly losing its emission despite its five thousand hours of flawless performance, a resistor is beginning to crack, or a tiny inductor is heading inexorably for an open-circuit. In two, five, ten years – nobody can guess when – the first failure will occur and a repair ship will have to set out to replace the faulty repeater. But with any luck, the first breakdown will not occur until the projected second telephone system has been built. When that happens, there may be delays, but the chance of a complete interruption of service will be vanishingly small.

CHAPTER 23
Laying the Cable

The attitude of the British Post Office to the telephone has changed somewhat since the day when its Engineer-in-Chief decided that Mr. Graham Bell's new invention was only of very limited use. It pioneered, during the late thirties, the development of submerged repeaters, and laid the first one in the Irish Sea in 1943. Many more were laid between England and the Continent in subsequent years, culminating in a 350-mile link with Norway in 1954. This circuit, the longest submarine telephone cable in the world until the laying of the transatlantic cable, contained seven deep-water repeaters which the Post Office had developed with an eye on the Atlantic circuit, and which in the actual event were used on the Newfoundland–Nova Scotia leg of the system. They could have been used for the main ocean crossing, if the problem of laying them without stopping the ship could have been solved in time.

The Post Office also provided the vessel and the know-how without which the cable could never have been safely strung from continent to continent. It is rather hard for those to whom the initials GPO conjure up red pillar-boxes and telephone kiosks to associate the Post Office with anything nautical, but it has been a shipowner since 1870, when it acquired the first *Monarch*, a small paddle-boat of some five hundred tons.

There have been three successively larger *Monarchs* since then; the last was sunk by enemy action in the closing weeks of the war – after having survived, soon after D-Day, an unfortunate mistake by an Allied destroyer which had cost her her captain, her bridge, and several of her crew.

Cable ships need a good deal of luck to survive in wartime, as when they are paying out they have to steam at a steady rate on an unvarying course. Besides the *Monarch*, the Post Office lost the smaller *Alert* under tragic circumstances. She was laying a cable to France in the wake of the Allied forces and was on the point of making the final splice when the electricians at the shore station noticed that the line was dead. Search vessels were sent out; they found the little ship's masts just above the water, still flying the international signals that denote cable-laying operations. She had been torpedoed, and there was not a single survivor.

The present *Monarch* was built in 1946 to offset these wartime losses, and with her tonnage of 8,050 she is the largest cable ship in the world, being able to carry 2,600 miles of deep-sea submarine cable. As a cable ship is the only vessel that progressively loses her cargo during the course of a voyage, her design has to be somewhat unusual. Much of the *Monarch*'s volume consists of four huge circular wells or tanks, forty-one feet in diameter, each of which can hold more than a thousand tons of cable. The coiling of such a mass of heavily armoured cable so that it can be paid out smoothly without kinks or tangles at up to ten miles an hour is an art in itself – as anyone who has ever engaged in a life-and-death struggle with a demoniacally possessed hosepipe will agree.

Before she became engaged in the transatlantic telephone project, the *Monarch* had already laid many important cables of very varied types in different parts of the world, not only for Britain but also for other countries. And not all of these cables were designed for communications: some were for the transmission of electric power – a field which has recently become of considerable importance with the announcement of plans to link the British and European power networks. Perhaps the most interesting of the *Monarch*'s off-trail jobs was the laying of the 1,500-mile submarine cable from the Air Force Missile Test Centre, Patrick Base, Florida, to the tracking stations above which so many strange shapes have roared out into the South Atlantic during the last few years. When the United States earth satellites were launched, the data on their initial trajectories flowed back along the cable that the *Monarch* laid from, as a GPO report put it, 'the most outlandish places, rarely used by the shipping of any nations'. And, one might add, even more rarely used these days, except by inquisitive Russian submarines.*

* Another Post Office publication concerning the *Monarch* concludes with this almost desperate note: '*Monarch* does not require a mascot. When she was launched over two hundred black cats were offered to the captain.'

Since the pioneering days of the *Great Eastern*, the hazards and problems of cable-laying have been immensely reduced by a whole series of scientific inventions which would have delighted the heart of Lord Kelvin. His deep-sea sounding device has been largely (but not entirely) superseded by that precursor of radar, the echo-sounder. Today, a ship's captain can obtain a sense of security never imagined in an earlier age as he watches the profile of the sea-bed being drawn on the chart, with such accuracy that all changes of level – and even large wrecks – are revealed at once. For a cable ship, few instruments are more valuable than the echo-sounder.

Radar itself is also an exceptionally useful aid, quite apart from its normal value in navigation. At night or in bad weather, it helps to locate buoys marking cable-ends, and gives a fix when a coastline is being approached. During the laying of the transatlantic telephone cable, where it was particularly important that the ship's position be known with great accuracy, a chain of Decca radio-navigation transmitters was set up in Newfoundland. These stations radiate waves which form a criss-crossing pattern on which a ship or aircraft can locate itself to within a few yards. There was no danger of the *Monarch* suffering the ignominious fate of some of her predecessors and running out of cable owing to errors in navigation. . . .

An operation as novel – and as expensive – as the laying of two thousand miles of repeatered telephone cable across the Atlantic required a full-scale dress rehearsal, which took place off Spain in the spring of 1955. Because the flexible repeaters for the deep-sea section of the cable could not be allowed to bend in a curve of less than three and a half feet radius, most of the *Monarch*'s paying-out gear had been extensively modified. This had to be tested, the crew had to be trained in its use, and measurements had to be made of the cable and repeater characteristics to see if the laboratory predictions were correct.

Shallow-water trials of the British rigid repeaters were made in water about a thousand feet deep forty miles west of Cadiz, and deep-water tests of the flexible American repeaters were carried out in the open Atlantic at depths of almost three miles. Forty miles of cable were paid out in a loop and transmission tests were made under what were virtually operating conditions.

When all the trials had been successfully concluded, everything was ready for the big job. It would have been simplest if the 2,200 miles of transatlantic cable could have been laid in a single operation, but the presence of the repeaters limited the amount of cable that the *Monarch* could carry. It was therefore necessary to do the laying in stages, and this involved a considerable amount of steaming back and forth across the Atlantic.

On June 28, 1955, the *Monarch* headed out from the Newfoundland coast with the first two-hundred-mile section of cable aboard. This length, rather more heavily armoured than the deep-sea section but electrically identical, led to the edge of the continental shelf, and its end was marked by a large buoy so that it could be easily located. There was a certain amount of excitement on this two-day mission when some icebergs were encountered on the cable route, but it was possible to avoid them without difficulty.

The *Monarch* then steamed back to England to pick up the main length of deep-sea cable from the Erith factory, and was back at the site of the cable-buoy in mid-August. Unfortunately, the chain attached to the cable had parted, and three days were spent grappling for the lost end – part of the time in a stiff gale. But this sort of unpleasantness has been standard operating procedure for cable ships for a hundred years.

What was not standard procedure, however, was the laying of the flexible repeaters, which involved slowing the ship's speed from six to two knots while they were going through the paying-out gear and stern sheaves. It must often have given the engineers and crew a queasy feeling to watch the bulges in the cable each costing as much as ten Cadillacs, bending in improbably sharp curves as they came out of the tanks, and not a few backbones must have twitched in sympathy.

Perhaps the most difficult problem which had to be overcome in the laying operations was the placing of the equalisers. These had to be inserted into the cable at intervals of several hundred miles to correct for slight and unpredictable variations in its electrical behaviour after it had reached the sea-bed; they were, in effect, fine adjustments to ensure that the delicate balance of cable loss versus amplifier gain was precisely kept.

Until the cable had been laid and measurements made on it,

there was no way of telling precisely where these equalisers had to be inserted. Under unfavourable conditions it could take half a day to make the two splices involved, yet the ship could not be stopped while the job was done, because of the danger that the cable might kink. The engineers had, therefore, to 'guess ahead', on the basis of their instrument readings, where an equaliser should be inserted. It might be sixty or seventy miles farther along the line, in the cable still waiting to be laid, and the jointing crews had to have everything ready by the time the equaliser was due to go overboard. A whole section of cable might have been lost had it become necessary to stop the ship.

Splicing a deep-sea cable is not a simple, straightforward job like soldering a couple of wires together, and in this case, where absolute reliability was vital, it was a major operation. The conductors had to be cleaned and brazed together, a length of the polythene insulator had to be injected into place, and all the armouring had to be rebuilt layer by layer. Every joint was X-rayed and had to be done again if there was the slightest defect. It is not surprising that it could take up to twelve hours to finish the job – especially if the weather was unco-operative.

On the whole, it was not; to quote from a subsequent (and probably autobiographical) report in the *Bell System Technical Journal*, 'generally speaking, the effect of the weather on the engineering supernumeraries on board was not severe, although *Monarch*'s stock of dramamine was somewhat depleted by the end of the project'.

The second length of the No. 1 (west to east) cable came to an end about 550 miles off Scotland, at the Rockall Bank. Rockall itself is a single barren spike sticking out of the sea, on which not more than a dozen men have ever landed. It is a sheer crag, often entirely covered with spray and almost always surrounded by a boiling surf which makes it unapproachable. Soon after the last war, in an empire-building mood, the Royal Navy landed a survey party from a helicopter and formally annexed it, to the utter indifference of the seagulls who are its only inhabitants.

It is not surprising that when she got back to Rockall Bank with the last section of the cable in her tanks, the *Monarch* discovered no sign of her buoy, which was later spotted more than five hundred miles away, apparently headed towards the North

Pole. So once again she had to grapple, and was in the process of doing this when she was hit by the high seas and 100 m.p.h. winds from Hurricane Ione, which drove her many miles off course and held up operations for several days. But finally the cable was located, the last 550 miles spliced in, and on September 26, 1955 the circuit to Scotland was completed. Newfoundland could speak to Europe by telephone, but Europe could not yet reply.

There is a considerable contrast between the two towns which are now invisibly linked by the two thousand miles of the Atlantic telephone. Clarenville is a small settlement of some 1,500 people, eight miles south of the great airport at Gander. It is a key point in the narrow-gauge Newfoundland railroad system, and though it might also be expected to be a fishing centre (as indeed it was in its early days) very few boats ply its waters now. This is one reason why it was chosen; the prospect of even a few anchors dragging on the sea-bed would have scared the cable engineers elsewhere.

In contrast, Oban is a much more substantial place, being a busy resort town during the summer months. It is set in one of the most beautiful parts of Scotland, with a couple of castles near by to give an additional attraction. But to describe the buildings lining its waterfront as 'medieval', as was done in one American Telephone and Telegraph Company publication, is a slight example of transatlantic hyperbole. Oban's waterfront is all too obviously Late Victorian neo-Gothic, though to confuse the architectural issue thoroughly the hill above the town is surmounted by a Roman Coliseum – the somewhat surprising family memorial of a local banker.

As cable-laying can be carried out in the open Atlantic only during the summer months (and not always then) the *Monarch* went back for her annual refit after completing the first cable. She set out again on April 18, 1956, carrying the three hundred miles of cable for the Newfoundland–Nova Scotia (Terrenceville to Sydney Mines) link of the chain. This laying was uneventful, the ship being stopped and the cable by-passed from the laying machinery whenever one of the 1,200-pound rigid repeaters emerged from the tank. The job was finished by early May, and *Monarch* headed back to England to start on her final task – the laying of the east-to-west No. 2 cable across the Atlantic.

An exact reverse of the earlier operation, this again involved laying three sections and splicing them in mid-ocean at Rockall Bank and again two hundred miles off Newfoundland. This time the weather was much more favourable, and the final splice was made on August 14, 1956 – ninety years and eighteen days after the *Great Eastern* had landed the end of the first successful telegraph cable in the same waters.

It cannot be said that there was any breathless moment of suspense when the last splice was made and the engineers knew whether the multi-million-dollar project on which they had laboured for years was successful or not. All the time that the *Monarch* had been paying out, except during the intervals when joints were being made, she had been able to talk along the cable to the shore stations. No one doubted that, barring accidents of the kind that can never be wholly guarded against at sea, the cable could be made to work, and work well. Once the circuits had been tested, the only people to be in any suspense were the Post Office and AT & T accountants, who had to satisfy themselves that the system would pay its way.

The official opening of the cable took place with appropriate ceremony on September 25, 1956. It was 11 a.m. when the chairman of the board of AT & T picked up the phone and said: 'This is Cleo Craig in New York calling Dr. Hill in London.'

There was a nerve-racking pause that seemed much longer than it really was. Then from the other side of the Atlantic, where it was already four o'clock in the afternoon, the Postmaster-General replied in the orotund tones with which, in his earlier incarnation as the 'Radio Doctor', he had sent so many Britons rushing to the medicine chest as he discoursed on embarrassing ailments, frequently at the most inappropriate times. 'Is that you, Mr. Craig? This is Dr. Hill in London. I'm delighted to hear your voice.'

Within a few hours, one of the twentieth century's greatest technical achievements passed into everyday use, with scarcely a second thought from those who were employing it. When the first member of the public had completed his call, he was asked what he thought of the new service. He was quite surprised to discover that he hadn't been talking by radio, and had no idea that his voice had been travelling by submarine cable.

The hundreds of engineers, administrators, technicians and seamen who had made the miracle possible were not unduly upset. They were already planning the second and better cable.

CHAPTER 24

The New Cables

The immediate success of the 1956 Transatlantic Telephone Cable – now known as TAT 1 – started an explosion of cable-laying across the oceans of the world. Yet, ironically, TAT 1 was obsolete even before it went into service, thanks to the invention of the transistor nine years before. As an achievement in communications engineering, it may be regarded as the last triumphant swan-song of the vacuum tube, which had served mankind so well for half a century.

In view of the revolutionary improvements that transistors have now brought to radios, hearing aids and all other types of electronic equipment, it may seem surprising that they were not employed in the first cable. They would have occupied only a fraction of the space needed by the system's 306 vacuum tubes, would have required very much less power – and no high-voltage supply at all.

Since the transistor was a Bell Labs invention* which had scarcely been hidden under a bushel, there must have been a very good reason for this policy. It was, in fact, inspired by the extreme conservatism which underlay the whole approach to the trans-atlantic telephone system. The vacuum tubes used had been tested for almost twenty years, and their characteristics were thoroughly known. But when the cable was being designed, the transistor was scarcely five years from the laboratory; though it was a healthy infant, no one could be quite certain how it would face an adult task like this. Some of the early models had failed after short periods of operation, having been 'poisoned' by minute quantities of water or other impurities, and there was no guarantee that defects might not develop in five, ten, fifteen . . . twenty years. So while the scientists struggled to de-bug this wonderful new invention – with what success we all know – the engineers continued to plan and build cables based on the vacuum tube

* The name 'transistor', incidentally, was devised by John Pierce.

technology which they were virtually sure would soon be out-moded.

An important *mechanical* development also entered the picture. In 1957, the British Post Office carried out pioneering experiments with a submarine cable in which the armouring which had been a standard feature for more than a hundred years was abandoned, and all the strength was provided by a steel wire at the *centre of the cable.* The only external protection was a tube of tough plastic, which is all that is needed in the calm of the ocean depths.

This new, light-weight cable puts much less strain on the laying machinery, and is thus far easier to handle and pay out than the armoured variety it replaces. Most important of all, it has no tendency to twist and kink; because of the spiral armouring, the older cables often rotated hundreds of times on their way to the sea-bed, with the unfortunate results mentioned in Chapter 8. The plastic-covered cable, on the other hand, is wholly free from torsion, and uncoils almost as easily as a length of string.

With such a cable, there is no danger of disastrous kinks developing if bad weather holds up laying. The cable need no longer be paid out in one continuous operation, so the ship can be stopped in deep water to splice and lower the bulky, rigid re-peaters which – despite the allure of their far greater capacity – could not be used in the first telephone cable. And so a purely mechanical development has had far-reaching communications consequences.

Mention must also be made of a very ingenious electronic device which more than doubled the capacity of TAT 1 soon after it went into service, and has since been applied to all the other cables. This is 'Time-Assignment Speech Interpolation' – shortened, needless to say, to TASI.

TASI depends upon the fact that a great deal (surprisingly, over 60 per cent) of ordinary conversation actually consists of silences. A sufficiently swift-acting device could listen out for these pauses, and take advantage of them by instantly switching in another conversation.

An analogy from ordinary life may be helpful here. A telephone cable is rather like a multi-laned highway, the separate syllables of speech being dotted along it like individual automobiles. Seen from the air, more than 50 per cent of even the most crowded

27. Technicians place a Telstar satellite model in a special thermo-vacuum chamber for tests to determine the thermal effects of sunlight on the satellite. The chamber is at Bell Laboratories' location in Whippany, New Jersey.

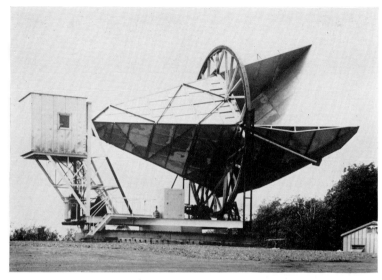

Above: 28. Horn antenna at Bell Laboratories, Holmdel, New Jersey. Originally built for the now famous NASA ECHO 1 experiment, the horn equipment was then modified to work with the Telstar communications satellite frequencies. The Holmdel horn antenna tracked Telstar and received broadband signals from the satellite.

Below: 29. Cut-away view of Bell System's experimental communications satellite, Telstar, launched in 1962.

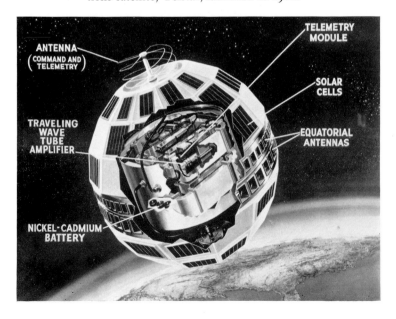

highway is empty space; its capacity could be doubled or tripled if vehicles could instantly 'lane-hop' to plug any gaps as soon as they appeared. Unfortunately, the laws of inertia, not to mention a few other difficulties, rule out such a happy solution to the traffic problem. But electric impulses, which have no inertia, and, equally important, all travel at exactly the same speed, can perform this trick.

As a result, the first 36-channel TAT 1 had its capacity increased almost at once to about a hundred channels. Later, 96-channel cables were able to carry no less than 235 simultaneous conversations, jumping from circuit to circuit every second or so. The mind boggles at all this simultaneous channel-hopping; presumably if, by bad luck, all 235 subscribers spoke at *precisely* the same moment, TASI would blow its fuses. However, the laws of probability indicate that such a catastrophe would not happen even once, in a universe as short-lived as ours.

By the end of the first year's operation of TAT 1, telephone traffic between the United States and England had doubled. A second cable, TAT 2, of identical design, went into service in 1959; however, it followed a different route, linking not Newfoundland and Scotland but Newfoundland and France.

A cable from Scotland to Canada, CANTAT 1, started operating in 1961 and was based on the new, non-twisting cable. It was now possible to have eighty circuits in a single cable, instead of only thirty-six in two cables. This immediately halved both the cost of laying, and the risk of damage.

TAT 3, in 1963, was the first cable direct from England to the United States: despite the greater distance involved, it was able to carry 138 circuits. TAT 4 (1965) provided a similar service between France and the United States.

TAT 5, which began operating in 1970, represented a real quantum jump in performance. The benefits of the solid-state revolution started by the transistor were now being realised; TAT 5 was able to carry 845 channels – a far cry from the original 36 of TAT 1, only fourteen years earlier. This cable used a new, southern route, from Spain to the United States. And four years later, in 1974, it in turn was eclipsed by CANTAT 2 – 1,840 channels! Two years after that, TAT 6 is designed to carry 4,000 circuits between France and the United States (see Table 1).

N

The Atlantic was not, of course, the only ocean where the cable ships were plying. A few months after the opening of TAT 1, a project of almost comparable magnitude was completed on the other side of the United States, when the Washington–Alaska cable went into operation. This 1,250-mile link from Seattle to Skagway could have been established entirely overland, but the reliability of the submerged repeaters was considered so good that the sea route was preferred. The circuits had to pass through territory where underwater conditions were frequently much less

TABLE I

Principal Atlantic and Pacific Telephone Cables

Name	Operating	Terminals	No. of Channels
TAT 1	1956	Scotland–Canada	36
TAT 2	1959	France–Canada	48
CANTAT 1	1961	Scotland–Canada	80
COMPAC	1963	Canada–Australia	80
TAT 3	1963	England–US	138
TRANSPAC	1964	Hawaii–Japan	138
TAT 4	1965	France–US	128
SEACOM	1965	Australia–Singapore	160
TAT 5	1970	Spain–US	845
CANTAT 2	1974	England-Canada	1,840
TAT 6	1976	France–US	4,000

* Note: TASI equipment can approximately double the number of available circuits on some cables.

unpleasant – and more equable – than those on land. The thirty-nine repeaters used in the twin cables spanning the main submarine section of the route were identical with those employed on the transatlantic circuit.

No sooner had the Alaska service been opened than work started on a project even more ambitious than the original Atlantic cable. This was the laying of what may be the longest submarine telephone link that will ever be constructed – the 2,400-mile 'Pacific Voiceway' between California and Hawaii.

Once again, *Monarch* played the leading role in the operation, but this time she was supported by another British cable ship, Submarine Cables Limited's *Ocean Layer*. Between them they laid 114 submerged repeaters in waters up to three miles deep, and the survey carried out to find the best route for the twin cables helped to fill in some of the blanks in the world's greatest ocean. An uncharted peak towering two miles above the sea-bed was discovered at one stage of the survey, and the cables were laid through the picturesquely named 'Moonless Mountains', a ten-thousand-foot-high range that runs from north to south for a thousand miles between California and Hawaii. When this second ocean cable was opened on October 8, 1957, it became possible to speak by cable from one side of the earth to the other.

But just four days earlier, on October 4, an event had occurred which would transform not only the future of global telecommunications – but almost everything else on the planet Earth.

CHAPTER 25

Communications Satellites

When the first edition of this book was written, soon after Sputnik I came up like thunder out of Kapustin Yar, it concluded with these words:

> It may well be that the submarine cable, even in the moment of its greatest technical triumph for a hundred years, is already doomed. . . . Even if this is so, there can be no doubt that it has still many decades of service ahead of it. It may not celebrate its second century, but nevertheless its old age will be even more vigorous and active than its youth . . . .

This has proved to be the case. The advent of communications satellites has so far been a stimulus, rather than a death-blow, to submarine cables – which, as Table 1 shows, improved their performance almost out of recognition within fifteen years. Something rather similar occurred with atomic and coal-fuelled power stations, during very nearly the same period of time. The 'Fossil Fuel' systems increased their efficiency so much that they were able to give stiff competition to the new source of energy.

The first suggestion that artificial satellites of the Earth might be used for communications appears in Chapter 20 ('Stations in Interplanetary Space') of Hermann Oberth's classic book *Wege zur Raumschiffahrt* (*Ways to Spaceflight*). This astonishing volume, published in 1928, outlines most of the basic principles of space-travel, as well as the reasons – scientific, commercial, military – for such an enterprise. Oberth pointed out that observers on space-stations 'with suitable reflectors, could send light signals to Earth. They make possible [*sic*] telegraphic communication with places that are cut off from normal contact by telegraphy because of operational disturbances' (NASA Technical Translation TT F-622). The same paragraph, incidentally, mentions the value of space-stations for Earth resources surveys, military reconnaissance, iceberg warning and meteorology. Seldom can such daring predictions have been so amply fulfilled, during the lifetime of the prophet.

It is curious that Oberth did not mention radio in this passage, but spoke only of light signals from 'small plane reflectors'. However, at that time long-range radio employed enormous antenna arrays, miles in extent (see Chapter 16) and the breakthrough into short waves was only just beginning. Difficult though it is for us to imagine such a state of affairs, in those days radio may not have seemed very practical for *really* long-range communications!

In the winter of 1944, as a Royal Air Force officer working on GCA (Ground Controlled Approach) radar – the so-called 'talkdown' system* – I was dealing every day with radio waves only about an inch in length, which could be generated at high power levels and focused into very narrow beams. I was also involved with my fellow enthusiasts in trying to arouse interest in spacetravel, specifically by reviving the British Interplanetary Society, which had been formed in 1933 but had been in suspended animation since the outbreak of hostilities. Was there any way, we asked ourselves with some desperation, in which rockets could be used to earn money – and thus, eventually, pay for spaceships? (Luckily for us, in those innocent days we had no idea just how expensive spaceships would prove to be.)

I had never heard of Oberth's proposal and though I acquired a photo-copy of *Wege zur Raumschiffahrt* (since autographed to me by the author) in 1945, I am ashamed to admit that I never read it until the NASA translation appeared in 1972. Though I do not recall any blinding moment of inspiration, sometime in late 1944 it became obvious to me that orbiting Earth satellites could provide a complete answer to most of the problems of global communications. Moreover, there was one particular orbit that the Almighty had clearly designed for the job. This was the so-called synchronous or geo-stationary orbit, 22,000 miles above the Equator; a satellite injected into such an orbit would appear to hover motionless high above the Earth, as if supported on an invisible tower.

The arrival of the first V2 rockets on London in late 1944 gave

* A somewhat dramatised, but basically accurate, account of these events will be found in my novel *Glide Path*.

me an opportunity to promote this idea. In a short letter printed
in the February 1945 issue of the British magazine *Wireless World*
I pointed out the great value of rockets for upper atmosphere
research and mentioned their possible use for launching instru-
mented satellites. The letter concluded:

> I would like to close by mentioning a possibility of the more
> remote future – perhaps half a century ahead. An 'artificial
> satellite' at the correct distance from the Earth would make one
> revolution every 24 hours; i.e. it would remain stationary above
> the same spot and would be within optical range of nearly half
> the earth's surface. Three repeater stations, 120 degrees apart in
> the correct orbit, could give television and microwave coverage
> to the entire planet.

Today, that phrase 'the more remote future – perhaps half a
century ahead' (i.e. 1995!) may raise some smiles. My excessive
conservatism illustrates the problem of setting a time-scale for any
new technological development. Based on my experience of wartime radar, I had assumed that the repeater stations would have to
be *manned*; it seemed obvious that engineering crews would be
needed to maintain and repair the masses of complex electronic
equipment that would be required for such a project. No one
could reasonably have anticipated the break-throughs in mini-
aturisation and reliability that made communications satellites
possible with payloads of a few hundred pounds. Meanwhile,
1995 remains a plausible target date for the large, permanently
manned space-stations which will grow from 1973's Skylab.

During the spring of 1945 I expanded these ideas and summed
them up in a four-page, single-spaced memorandum dated May
25. Three or four copies of this carefully typed document were
circulated among the officers of the then renaissant British
Interplanetary Society. Some twenty years later, when I assumed
that they had all been lost, Ralph Slazenger (of the sporting goods
firm) discovered the top copy, in absolutely mint condition, in his
files. I was promptly relieved of it by my good friend Fred Durant,
Assistant Director (Astronautics) of the National Air and Space
Museum, so it is now safely preserved in the Smithsonian.

Soon afterwards, I re-wrote this memorandum in the form of
an article which I submitted to *Wireless World*. My original title

was a forthright 'The Future of World Communications', but when it was published in October 1945 the editor changed it to a less grandiose but more informative 'Extra-Terrestrial Relays'.

This paper* was the first public presentation of the communications satellite concept, and during the next decade the idea became familiar to all those interested in space-travel. But nothing much could be done about it until the next generation of launch vehicles was ready, at the end of the 1950s. After that, progress was almost unbelievably swift.

The very first communications satellite was the Project SCORE Atlas of December 1958, which broadcast to the world a pre-recorded Christmas greeting from President Eisenhower. It also received and re-transmitted a number of voice and teletype messages. About a year later (October 1960) the US Army launched a much more complex satellite, COURIER 1B, but it failed after eighteen days, though not before it had transmitted more than a hundred million words.

Meanwhile, however, the eyes of the world had been, quite literally, focused upon another type of communications satellite. The 100-foot diameter balloon ECHO 1 was launched on August 12, 1960, and was visible to millions of people as a brilliant star moving slowly across the night sky. The ECHO project was to a large extent the brainchild of Dr. John Pierce, the instigator of this book; he has described his involvement in his monograph *The Beginnings of Satellite Communications* (San Francisco Press, 1968).

For many years John had been moonlighting as a science-fiction writer ('J. J. Coupling') from his job as Director of Communications Research at Bell Labs. In 1954 he combined both his interests in a paper entitled 'Orbital Radio Relays' (*Jet Propulsion*, April 1955). He had not then seen my 1945 article, but covered the same ground and came to very similar conclusions. Being a practical engineer, he was able to focus his eyes less than fifty years ahead, and pointed out that we did not have to go all the way up to the 22,000-mile-high stationary orbit to obtain some very useful results. A great deal could be done in quite a low orbit – say one or two thousand miles – and with a satellite which did not even

* Printed as an appendix to *Voices From the Sky*, which also contains several essays on the history and potential of communications satellites.

carry a transmitter. All that was required was a good reflector, such as a large silvered balloon which would act as a radio mirror. (This proposal was first made in a pioneering paper 'Minimum Satellite Vehicles' by K. W. Gatland, A. M. Kunesch and A. D. Dixon, published by the British Interplanetary Society in 1951.)

ECHO I, and its slightly larger successor ECHO 2 (1964) demonstrated the feasibility of this idea, but also its limitations. Since a spherical mirror scatters the radiation falling upon it in all directions, all but a minute fraction is wasted. To detect the signals from Echo it was necessary to employ huge and expensive antennae, backed by super-sensitive receivers. A practical communications system could not be based on such 'passive' reflectors; what was needed was an *active* transmitter in orbit, fitted with antennae which would beam its messages to any desired point. Such a system would be millions of times more efficient than Echo; its achievement took less than a decade.

The next major step was the famous Telstar (July 1962) which made possible the first spanning of the Atlantic by television. It was also the first privately owned satellite, since American Telephone and Telegraph financed it and hired a launch vehicle from NASA for $3,500,000. Telstar was designed by Bell Labs, with John Pierce again in a major role, and its operation depended upon three of the Laboratories' inventions; the transistor, solar cells (which convert sunlight into electricity) and the travelling wave tube. The TWT is a special type of vacuum tube, the brainchild of Rudolph Kompfner, an Austrian *architect* who taught himself physics while interned in England during the early part of World War II. (I am happy to say that before the war was over, he was working on classified microwave research for the Admiralty.) The TWT has a unique ability to amplify an enormously wide band of radio frequencies; it has played a key role in satellite communications – so much so that it has often been the only vacuum tube aboard some satellites. All the rest of the electronics has become solid state.

Telstar was such a spectacular success that it is now almost synonymous with communications satellites, to the mortification of AT & T's competitors and the annoyance of its lawyers (who had carefully registered the name). It quite overshadowed NASA's own communications satellite, Relay, built by RCA and launched later in the same year. But Relay was also a success, and its chief claim to fame was the largest TV audience in history up to that time, when it carried President Kennedy's funeral to the world.

Telstar and Relay, as well as their successors Telstar II and Relay II, led to the development of the now familiar ground stations, with their giant parabolic mirrors, which form the essential terrestrial component of any space communications system. These ground stations are now still in use, but the Telstars and Relays – though they will orbit for centuries – have long been silent. Once they had served their purpose, they were quickly replaced by more advanced satellites at much greater altitudes.

The trouble with the low-altitude satellites such as Telstar (maximum distance from the Earth, 6,700 miles) is that they move fairly rapidly across the sky and so can provide service, at any given station, for only short periods of time. To overcome this problem it is necessary to have a whole string of satellites, equally spaced around the earth, and to keep track of them all as they drift across the heavens. Even then, there will probably be breaks in communication, when one satellite drops down below the horizon before another comes up.

The alternative, of course, is to go up to the synchronous, 22,000-mile-high orbit, where the satellite stays fixed in the sky and a simple, *motionless* antenna, with no elaborate tracking equipment, can remain pointed towards it for ever. This solution seems so obvious (I never considered any other in my 1945 paper) that it may be wondered why anyone ever bothered about lower orbits.

There were several good reasons. In the first place it takes a great deal of energy to reach synchronous orbit – more, surprisingly, than to escape completely from the Earth! The launch vehicles available in the early 1960s could not carry a worthwhile payload to such a height, but with improved performance this difficulty was overcome well before the end of the decade.

A second problem was of a much more fundamental nature, and had never been encountered (at least in serious form) in terrestrial communications. Since radio waves travel at 186,000 miles a second, they can get from one side of the Earth to the other in less than a tenth of a second. But the trip up to synchronous orbit and back takes a quarter of a second, so if you are having a conversation with someone over such a link it will take twice this time, or half a second, before you can get a reply – even if your listener reacts instantly.

Half a second may not seem a serious delay, but it is quite perceptible. Moreover, it introduces the possibility of something much worse. In all long-distance circuits, there is danger of echo, caused by a small fraction of the power being reflected back to the speaker, who will thus hear his own voice after a slight delay. Special circuits ('echo suppressors') have been developed to deal with this phenomenon, which can be quite intolerable. It is literally impossible to talk, when you can hear your own voice a fraction of a second later; in fact, this is a good way of exposing anyone who is faking deafness.

It was believed by many that these twin problems – time delay, and echo – might rule out the use of synchronous satellites for telephone conversations. Of course, one-way radio and TV broadcasts would be unaffected, since the ordinary viewer or listener would neither know nor care if the transmission had taken seconds – or even minutes – to reach him.

Fortunately, it turned out that echo suppressors could be adapted for satellite links, and the unavoidable time-lag is little more than a barely perceptible nuisance – except for people who like to keep interrupting each other. What is even more surprising is the fact that the *five*-fold longer delay to the Moon (two and a half seconds for the round trip) causes so few problems. It is quite easy to overlook it, when listening to conversations between Mission Control and astronauts on the lunar surface. Admittedly, they have better radio discipline than, for example, a couple of Latin business men arguing over a deal.

Just as John Pierce was the gadfly largely responsible for involving Bell Labs in the ECHO and TELSTAR projects, so three young engineers at the Hughes Aircraft Company laid their careers on the line to promote the synchronous satellites upon which, they were convinced, the future of world communications depended.* Harold Rosen, Donald Williams and Thomas

* Though literally thousands of engineers and scientists contributed to the development of communications satellites, there are two more who deserve special mention here. One is the electronic engineer and science-

Hudspeth had little success in selling their ideas to the Hughes management until Williams walked into Vice President Lawrence Hyland's office one day with a cheque for $10,000 – his entire savings – as a contribution to the project. The donation was not accepted, but the company decided that such enthusiasm should be supported. The result was the Syncom series (1963–4) and finally Early Bird (1965), which may be regarded as the precursor of today's operational satellites.

Early Bird (or Intelsat 1, as it has now become) was the first-born of the newly created Communications Satellite Corporation, invariably shortened to COMSAT. And that in turn was the child of the 1962 Communications Satellite Act, drafted only five years after the beginning of the Space Age. By this time it was obvious that legalisation was needed to control, and encourage, the new mode of communications that was imminent, but just *how* it should be done was far from obvious.

Some said it should be left to private enterprise, and pointed to the US telephone system as a model. Others argued that this would involve a give-away of billions of dollars of government-sponsored research – and as vital matters of security and international relations would be involved, so great a power demanded some form of state control. After long and heated debate in Congress, COMSAT was set up as a private company, but with a considerable degree of supervision by the President, NASA and the Federal Communications Commission. Its fifteen-man Board of Directors was to have six members elected by the public (who owned half the stock), six elected by the national communications

---

fiction writer George O. Smith, who published the *Venus Equilateral* stories in 'Astounding Stories' during the early 1940s while he was working on that most improbable device, the radio proximity fuse. 'Venus Equilateral' was a relay station forming an equilateral triangle with Venus and the Sun, so that radio communications with Earth and that planet could always be maintained. I was very familiar with these stories and, as John Pierce has pointed out, they may have subconsciously influenced my thinking.

The other is Robert P. Haviland, of the General Electric Company, who in 1946 skilfully used my *Wireless World* paper as a carrot to coax the US Navy into the Space Age, and has since published many important studies of communications satellites.

companies (who owned the other half) and three appointed by the President, with the consent of the Senate.

The setting up of this monopoly did not please everyone. John Pierce, for example, was bitterly disappointed that AT & T could not follow up the success of Telstar and establish itself in space. In his memoir he remarks, 'I foresaw that the Act would, as it did, considerably delay the realisation of a commercial satellite system.' COMSAT would hardly agree.

When Early Bird started operations in June 1965, its 240 two-way voice channels increased the number of circuits across the Atlantic by over 50 per cent. Designed for eighteen months of service, it ran for forty-two with perfect reliability before its successors took over; even then, it was sometimes called back from retirement in emergencies – for example, to give extra coverage during the Apollo 11 mission in July 1969.

The swift establishment of the global system is best shown by table 2 on page 205.

The truly global system I described in 1945 first became reality in July 1969, when a TRW Intelsat III took up position over the Indian Ocean, linking up with identical satellites over the Atlantic and Pacific. Since that time there has been no spot on Earth (except for a small region round each Pole) beyond the reach of direct TV communications. Perhaps this was shown most dramatically during President Nixon's visit to mainland China, which before February 1972 seemed almost as remote and mysterious as the far side of the Moon. It is impossible to overestimate the impact of seeing, often in real time, great political events that affect the lives of millions. To give but one example; it was probably the TV coverage of the Vietnam War, more than any other factor, which eventually forced the United States to withdraw.

Starting in the late 1960's, more and more countries acquired ground stations so that they would be able to plug themselves into the rest of the global electronic village. Some kind of international agreement to operate the system was obviously essential, and after several years of complex negotiations the definitive arrangements were drawn up between the eighty member countries forming INTELSAT (the International Telecommunications Satellite Organisation). The agreement was signed on August 20, 1971, at an impressive ceremony at the State Department. As one of the

TABLE 2

The Intelsat Series

Satellite	Date & Maker	Capacity	Investment Cost per Circuit Year	Notes
Early Bird (Intelsat I)	Hughes 1965	240 Voice or 1 TV	$15,300	First Commercial Comsat
Intelsat II	Hughes 1966–7	240 Voice or 1 TV	$8,400	Three put into service (1 Atlantic 2 Pacific) Now in reserve
Intelsat III	TRW 1968–9	1,200 Voice or 4 TV	$1,450	Established global system (Atlantic, Pacific, Indian Ocean) in July 1969
Intelsat IV	Hughes 1971–2	6,000 Voice or 12 TV	$500	Four provided global service from June 1972. First satellite with steerable spot-beams.

speakers at the subsequent banquet, I would like to quote some of my remarks on that occasion:

Let me remind you that, whatever the history books say, this great country was created a little more than a hundred years ago by two inventions. Without them, the United States was impossible; with them, it was inevitable. Those inventions, of course, were the railroad and the electric telegraph.

Today, we are seeing on a global scale an almost exact parallel to that situation. What the railroads and the telegraph did here a century ago, the jets and the communications satellites are doing now to all the world.

I hope you will remember this analogy in the years ahead. For today, whether you intend it or not – whether you *wish* it or not – you have signed far more than just another inter-governmental agreement.

You have just signed a first draft of the Articles of Federation of the United States of Earth.

CHAPTER 26

Star of India

Although communications satellites were first employed to bridge the oceans, it was immediately obvious that they could be equally valuable in providing services to large land areas – particularly those which were undeveloped or sparsely populated, and did not possess the coaxial and microwave networks of Europe and the United States. It is not surprising that, with its lead in space technology, the USSR was the first country to develop a domestic system, based on its Molniya ('Lighting') satellites. Thirteen of these were launched between 1965 and 1970, and about forty ground stations were set up all over the USSR to operate through them. The Molniyas do not employ the twenty-four hour synchronous orbit, but an interesting compromise first suggested by a British space scientist, Dr. William Hilton. This is a twelve-hour, highly elliptical orbit, in which the satellite is 25,000 miles above the Earth at its highest point – over the northern hemisphere – and only 300 miles up at its lowest, over the southern hemisphere. The orbit is also steeply inclined (sixty-five degrees) to the Equator. As a result, the satellite hovers high above Russia for several hours twice a day – at exactly the same time every day – then plunges down behind the Earth and comes up again for the next period of service.

The first domestic *synchronous* system was established by Canada in 1972 with the Hughes-built Anik ('Brother') satellites, capable of providing 6,000 telephone circuits or twelve colour TV channels. The Canadians had realised that such a system was vital if they wished to develop the enormous resources of the Far North; people simply would not work or live in such remote regions unless they could talk to their friends, call doctors in emergencies, and see their favourite TV programmes.

There is considerable irony in the fact that Canada established a domestic satellite system before the United States – and, in fact, has leased some of its channels to US customers! But bitter in-fighting between the communications companies and the United States Government as to who should own and run such a system held up all progress for years; not until December 1972 was COMSAT finally given a go-ahead by the Federal Communication Commission. Some of the legal problems involved may be gathered from this extract from COMSAT's 1972 Annual Report, which I suggest you read slowly:

. . . the FCC removed some proposed restrictions on COMSAT and authorized it to proceed with one program for domestic US service and to have a one-third participation in another. The FCC order directed that COMSAT form a subsidiary to conduct all its domestic communications activities. In compliance, we have established *Comsat General Corporation*, which will be responsible for all new programs in which COMSAT may participate outside of the INTELSAT organisation.

In the first of our domestic programs, COMSAT General will lease satellites to American Telephone and Telegraph Company. Under the agreement, COMSAT General will provide, own and operate three satellites . . . each with a capacity of approximately 14,400 two-way telephone circuits . . . Earth Stations for communications service with the satellites will be provided, owned and operated by AT & T.

In its order, the FCC said that AT & T may not use the satellites for competitive services until all other domestic satellite systems are substantially utilised, or until at least three years after it begins domestic satellite operations. . . . In either circumstance . . . it must divest itself of COMSAT stock . . . As a step towards implementing a second domestic satellite program, COMSAT through COMSAT General has joined with MCI Communications Corporation and Lockheed Aircraft Corporation in a separate company to provide multipurpose US domestic satellite services to all customers except AT & T. COMSAT General, MCI and Lockheed each has a one third interest in the new company, CML Satellite Corporation (CML) . . .

This remarkable statement shows the legal contortions necessary because AT & T (the nation's largest communications carrier) is both COMSAT's chief customer and – through the submarine cables – its chief competitor for international services. All

Above: 30. AT and T's cable ship *Long Lines* passes the Statue of Liberty.

Below: 31. Underwater photograph shows deep-sea divers adjusting tow line on the sea plough. The plough, designed by Bell Laboratories to bury undersea telephone cable below the ocean floor, was used by the American Telephone and Telegraph Company's Long Lines Department to bury the shore-end sections of two transatlantic cables off the coast of New Jersey.

32. Model of Telstar II against artificial background.

too often, it seems, technical problems are simpler to solve than political ones. It took seven years to get this agreement from the FCC, a period which just about spanned the entire history of communications satellites. Perhaps John Pierce did have a point when he complained that the 1962 Act had caused 'considerable delay'.

Meanwhile, while the lawyers argue, the engineers push ahead. The satellites of the late 1970s will be able to carry tens of thousands of conversations simultaneously, or scores of TV channels. They will be able to switch their many individual and tightly focused beams from point to point on the Earth's surface, according to operational requirements – thus vastly increasing efficiency. (Intelsat IV was first to do this.) Equally important advances will have taken place on the Earth. Because the satellites can now be positioned accurately in the sky by means of small reaction jets fuelled for years of operation, it is no longer necessary to have complex tracking antennae at the ground stations. Fixed – and therefore far cheaper – reflectors can be used; and they can be relatively small, thanks to the increased power of the satellites. (As early as 1969, narrow-band signals from the military satellite TACSAT I (far left, Plate 24) could be received by antennae only *one foot* in diameter!)

With this increasing power and technical sophistication, a whole range of new possibilities is now opening up. The first communications satellites merely provided links between the already existing TV or telephone networks on the ground – but later satellites may replace, to a large extent, such terrestrial systems by allowing *direct* broadcasts from space to millions of small, individual receivers. This prospect is of enormous interest to all the developing countries, which do not possess, and could not afford, the cable and microwave networks which Europe and the United States have built. Just as they have often gone directly from ox-carts to aeroplanes, by-passing railroads, so many countries in Asia and Africa may leapfrog a whole era of communications technology and go straight to the Space Age. If all goes well (how many times one has heard that phrase in connection with space!) the new era will dawn in the country which needs it most – India.

In 1969 a joint agreement was signed between the United States and Indian Governments for the establishment, by NASA,

o

of the first TV broadcast satellite over the sub-continent. Applications Technology Satellite F (ATS-F) would be the first in the world powerful enough to be received *directly* by an ordinary domestic receiver, working in conjunction with a simple ten-foot-diameter wire-mesh reflector and pre-amplifier. This additional equipment, it was hoped, could be mass produced for between one and two hundred dollars.

ATS-F, built by the Fairchild Corporation of Germantown, Pennsylvania, would make possible this new level of capability by deploying in orbit the largest antenna ever carried by a satellite – a thirty-foot diameter parabolic reflector. This elegant structure would be folded up before launch, and would open up at the appropriate time rather like one of those Japanese artificial 'flowers' when dropped into water. (I have not seen one of these for years; perhaps the nimble fingers that used to make them are now more profitably employed assembling transistor radios.) The thirty-foot dish would be large enough to beam essentially all of the satellite's power on to India, thus giving a much higher concentration of energy than any other satellite had previously achieved.

The project was planned to start in the early 1970s, broadcasting first to schools and remote communities in the far west of the United States. Then the satellite would be nudged along the Equator to a point just south of India, for a year's operations there. The programmes would be beamed up from a large ground station near Ahmedabad, and re-broadcast so that they could be received over the whole of India.

Numerous administrative and technical delays postponed the target date; the most tragic setback was the death of its chief protagonist, India's able and energetic Dr. Vikram Sarabhai, who wore himself out as Chairman of his country's Atomic Energy *and* Space Commissions (and the International Atomic Energy Commission!). The most fitting memorial to Dr. Sarabhai would be the successful achievement of his dream; in any event, the philosophy behind it is equally applicable to all the developing nations of Africa, South America, Asia – who are watching the experiment with the greatest interest. So in what follows, you may substitute for 'India' the names of Brazil, Tanzania, Mexico, Indonesia . . .

It is sometimes quite difficult for those from nations which have taken a century and a half to slog from semaphore to satellite to appreciate that a few hundred pounds in orbit can now replace the continent-wide networks of microwave towers, coaxial cables and ground transmitters that have been constructed during the last generation. And it is perhaps even more difficult, to those who think of television exclusively in terms of old Hollywood movies, give-away contests and soap commercials to see any sense in spreading these boons to places which do not yet enjoy them. Almost *any* other use of the money, it might be argued, would be more beneficial. . . .

Such a reaction is typical of those who came from developed (or over-developed) countries, and who accept libraries, telephones, cinemas, radio, TV, as part of their daily lives. Because they frequently suffer from the modern scourge of information pollution, they cannot imagine its deadly opposite – information starvation. For any Westerner, however well meaning, to tell an Indian villager that he would be better off without access to the world's news, knowledge and entertainment is a gross impertinence. A fat man has as much right to preach the virtues of abstemiousness to the hungry.

Those who actually live in the East, and know its problems, are in the best position to appreciate what cheap and high-quality communications could do to improve standards of living and reduce social inequalities. Illiteracy, ignorance and superstition are not merely the results of poverty – they are part of its cause, forming a self-perpetuating system which has lasted for centuries, and which cannot be changed without fundamental advances in education. India's Satellite Instructional Television Experiment (SITE) is a bold attempt to harness the technology of space for this task; if it succeeds, the implications for all developing nations will be enormous.

SITE's first order of business will be instruction in family planning, upon which the future of India (and all other countries) now depends. Puppet shows are already being produced to put across the basic concepts; those of us who remember the traditional activities of Punch and Judy may find this idea faintly hilarious. However, there is probably no better way of reaching audiences who are unable to read, but who are familiar with the travelling

puppeteers who for generations have brought the sagas of Rama
and Sita and Hanuman into the villages.

Some officials have stated, perhaps optimistically, that the only
way in which India can check its population explosion is by mass
propaganda from satellite – which alone can project the unique
authority and impact of the TV set into every village in the land.
If this is true, we have a situation which should indeed give pause
to those who have criticised the billions spent on space. The
emerging countries of the so-called Third World may need
rockets and satellites much more desperately than the advanced
nations which built them. Swords into ploughshares is an obsolete
metaphor; we can now turn missiles into blackboards.

Next to family planning, India's greatest need is increased
agricultural productivity. This involves spreading information
about animal husbandry, new seeds, fertilisers, pesticides and so
forth; the ubiquitous transistor radio has already played an important role here. In some parts of the country, the famous 'Miracle
Rice' strains – which have unexpectedly given the whole of
Asia a few priceless years in which to avert famine – are known
as 'radio paddy', because of the medium through which farmers
were introduced to the new crops. But although radio can do a
great deal, it cannot match the effectiveness of television; and of
course there are many types of information that can be fully
conveyed only by images. Merely *telling* a farmer how to improve
his herds or harvest is seldom effective. But seeing is believing, if
he can compare the pictures on the screen with the scrawny
cattle and the dispirited crops around him.

Although the SITE project sounds very well on paper, only
actual experience will show if it works in practice. The 'hardware'
is straightforward and even conventional in terms of today's
satellite technology; it is the 'software' – the actual programmes –
that will determine the success or failure of the experiment. In
1967, a pilot project was started in eighty villages round New
Delhi, which were equipped with television receivers tuned to the
local station. (In striking contrast to a satellite transmitter, this
has a range of only twenty-five miles.) It was found that an average
of 400 villagers gathered at each of the evening 'Teleclubs', to
watch programmes on weed control, fertilisers, packaging, high-
yield seeds – plus five minutes of song and dance to sweeten the

educational pill. The TV viewers showed substantial gains of agricultural knowledge over the non-viewers; to quote from the survey carried out by Dr. Prasad Vepa of the Indian National Committee for Space Research:

> They expressed their opinion that the information given through these programs was more comprehensive and clearer compared to that of other mass media. Yet another reason cited for the utility of TV was its appeal to the illiterate and small farmers to *whom information somehow just does not trickle* [My italics].

In February 1971, while filming *The Promise of Space*, I visited one of these TV-equipped villages – Sultanpur, a prosperous and progressive community just outside Delhi, only a few miles from the soaring sandstone tower of the Kutb Minar. The Atomic Energy Commission had kindly loaned us a prototype of the ten-foot diameter, chicken-wire receiving dish which will collect signals from ATS-F as it hovers above the Equator. While the villagers watched, the pie-shaped pieces of the reflector were assembled – a job that can be performed by unskilled labour in a couple of hours. When it was finished, we had something that looked like a large aluminium sunshade or umbrella with a collecting antenna in place of the handle. As the whole assembly was tilted up at the sky and lifted on to the roof of the highest building, it looked as if a small flying saucer had swooped down upon Sultanpur.

With the Delhi transmitter standing-in for the still unlaunched satellite, we were able to show a preview of – hopefully – almost any Indian village of the 1980s. Yet it was not at Sultanpur, but 400 miles away at Ahmedabad, that I really began to appreciate what could be done through even the most elementary education at the village level.

Near Ahmedabad is the big fifty-foot diameter parabolic dish of the Experimental Satellite Communication Ground Station, through which the programmes will be beamed up to the hovering satellite. Also in this area is AMUL, the largest dairy co-operative in the world, to which more than a *quarter of a million* farmers belong. After we had finished filming at the big dish, our camera team drove out to the AMUL headquarters, and we accompanied the Chief Veterinary Officer on his rounds.

At our first stop, we ran into a moving little drama that we could never have contrived deliberately, and which summed up half the problems of India in a single episode. A buffalo calf was dying, watched over by a tearful old lady who now saw most of her worldly wealth about to disappear. If she had called the vet a few days before – there was a telephone in the village for this very purpose – he could easily have saved the calf. But she had tried charms and magic first; they are not *always* ineffective, but antibiotics are rather more reliable . . .

I will not quickly forget the haggard, tear-streaked face of that old lady in Gujerat; yet her example could be multiplied a million times. The loss of real wealth throughout India because of ignorance or superstition must be staggering. If it saved only a few calves per year, or increased productivity only a few per cent, the TV set in the village square would quickly pay for itself. The very capable men who run AMUL realise this; they are so impressed by the possibilities of TV education that they plan to build their own station to broadcast to their quarter of a million farmers. They have the money, and they cannot wait for the satellite – though it will reach an audience two thousand times larger, for over half a *billion* people will lie within range of ATS-F.

There is a less obvious, yet perhaps even more important, way in which the prosperity and sometimes very existence of the Indian villagers will one day depend upon space technology. The life of the sub-continent is dominated by the monsoon, which brings 80 per cent of the annual rainfall between June and September. The date of onset of the monsoon, however, can vary by several weeks – with disastrous results to the farmer, if he mistimes the planting of his crops.

Now, for the first time, the all-seeing eye of the meteorological satellites, feeding information to giant computers, gives real hope of dramatic improvements in weather forecasting. But forecasts will be no use unless they get to the farmers in their half a million scattered villages, and to quote from a recent Indian report:

> This cannot be achieved by conventional methods of telegrams and wireless broadcasts. Only a space communications

system employing TV will be . . . able to provide the farmer with something like a personal briefing . . . such a nation-wide rural TV broadcast system can be expected to effect an increased agricultural production of at least 10 % through the prevention of losses – a saving of $1,600 million per annum.

Even if this figure is wildly optimistic, it appears that the costs of such a system would be negligible compared to its benefits.

And those who are unimpressed by mere dollars should also consider the human aspect – as demonstrated, for example, by the great cyclone that hit Bangladesh in 1971. *That* was tracked by the weather satellites – but the warning network that might have saved half a million lives did not exist. Such tragedies will be impossible in a world of efficient space communications.

Yet it is the quality, not the quantity, of life that really matters. Men need information, news, mental stimulus, entertainment. For the first time in five thousand years a technology now exists which can halt and perhaps even reverse the flow from the country to the city. The communications satellite can put an end to cultural deprivation caused by geography. It is strange to think that, in the long run, the cure for Calcutta (not to mention London, New York, Tokyo . . .) may lie 22,000 miles out in space.

The SITE project will run for one year, and will broadcast to about five thousand TV sets in carefully selected areas. This figure may not seem impressive when one considers the size of India, but it requires only one receiver per village to start a social, economic and educational revolution. If the experiment is a success, then the next step would be for India to have a full-time communications satellite of her own. This is, in any case, essential for the country's internal radio, telegraph, telephone and telex services.

It may well be that, until it has established such a nation-wide system, India will be unable to achieve a real cultural identity, but will remain merely a collection of states. And one may wonder how much bloodshed and misery might have been avoided, had the two severed wings of Pakistan been able to talk to each other face to face, through the facilities which only a communications satellite could have provided.

Kipling, who wrote a story about 'wireless' and a poem to the deep-sea cables, would have been delighted by the electronic dawn that is about to break upon the sub-continent. Gandhi, on the other hand, would probably have been less enthusiastic; for much of the India that he knew will not survive the changes that are now coming.

One of the most magical moments of Satyajit Ray's exquisite *Pather Panchali* is when the little boy Apu hears for the first time the Aeolian music of the telegraph wires on the windy plain. Soon those singing wires will have gone for ever; but a new generation of Apus will be watching, wide-eyed, when the science of a later age draws down pictures from the sky – and opens up for all the children of India a window on the world.

Epilogue

There is no such thing as finality in this field, as in any other. My Lords of the Admiralty, who were not stupid men, were unable to see in 1820 that Francis Ronald's electric telegraph was any improvement on the clumsy semaphore system the Navy used to keep Portsmouth in touch with Whitehall. Nor was it much of an improvement – *if* one was concerned only with announcing the arrival and departure of ships. What no one could appreciate at the time was that the telegraph would open up an entire new dimension in communications and would transform business, social life, politics and news reporting, so that the pre-telegraph age would soon appear antediluvian.

The spread of radio and TV in our age is no more than a continuation of this process; the laying of the submarine cables, and launching of communications satellites, the essential linking-together of the various continental networks. Human society is like a living organism which is gradually evolving its nervous system, so that it can keep in touch with all its members, and know what every part of itself is doing, no matter where it may be.

We are still very far from achieving this, but it is the goal towards which all progress in communication consciously or unconsciously leads. It may be attained by means as unimaginable today as our electronic devices would have been to any man born a hundred years ago, but sooner or later it will be achieved, and the last barriers of distance will be down.

Scientists do not usually indulge in prophecy, and in these pages we have had many examples of eminent men who could not even see facts that were staring them in the face. It is a welcome change, therefore, to close this book with an extraordinary prediction made by the electrical engineer Professor Ayrton in the closing years of the nineteenth century. It is extraordinary because to the professor's listeners at the Imperial Institute on February 15, 1897, it must have appeared pure fantasy, and even today, with

o*

our far greater knowledge, it represents something still completely beyond attainment.

Ayrton had been lecturing on submarine telegraphy and ended his speech thus:

> There is no doubt that the day will come, maybe when you and I are forgotten, when copper wires, gutta-percha coverings and iron sheathings will be relegated to the Museum of Antiquities. Then, when a person wants to telegraph to a friend, he knows not where, he will call in an electro-magnetic voice, which will be heard loud by him who has the electro-magnetic ear, but will be silent to everyone else. He will call 'Where are you?' and the reply will come 'I am at the bottom of the coal mine' or 'Crossing the Andes' or 'In the middle of the Pacific'; or perhaps no reply will come at all, and he may then conclude that his friend is dead.

This prediction was quoted somewhat sarcastically at the time as an example of what happens when a scientist lets his imagination run away with him. But it was an inspired glimpse of a future which some of us may live to see – a future when, whether they like it or not, for better or for worse, all the men on earth are neighbours.

Acknowledgements

This book could not have been written without the co-operation of the Bell Telephone Laboratories and the American Telephone and Telegraph Company, and in particular of Dr. John R. Pierce, Director of Electronic Research at the Bell Labs, who kept air-mailing small mountains of literature to me in Ceylon.

My thanks for information and help are also due to Mr. G. R. M. Garrett, of the Science Museum, South Kensington, London, who made available to me many of the illustrations from his valuable booklet *One Hundred Years of Submarine Cables* (H.M. Stationery Office).

At the General Post Office, I would like to acknowledge the help of Mr. S. A. Manser, Deputy Director, External Telecommunications Executive; Mr. R. J. Halsey, Assistant Engineer-in-Chief; Captain W. H. Leech, O.B.E., D.S.C., Submarine Superintendent; Mr. Sidney R. Campion, Principal Information Officer.

Mr. H. J. Wilson, Public Relations Officer of Cable and Wireless Ltd., and Mr. P. L. Reed, Public Relations Officer of the Telegraph Construction and Maintenance Company and Submarine Cables Ltd., also gave me valuable assistance and provided illustrations.

Finally, it is a particular pleasure to record my thanks to the Institution of Electrical Engineers, Savoy Place, London, for the use of its fine library and the helpful co-operation of its staff.

London–Colombo–New York

Index

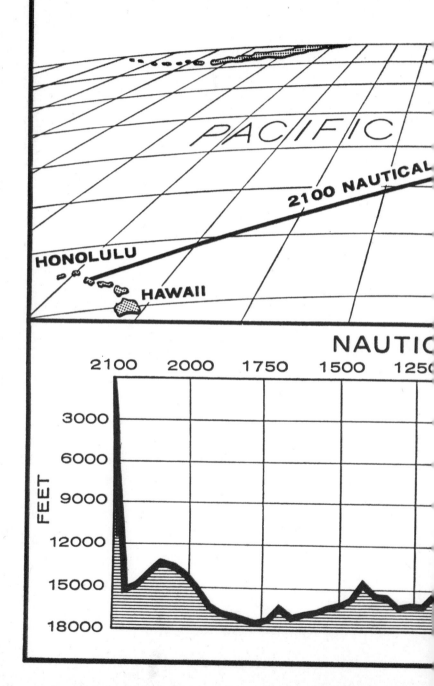